Who's Minding the Farm?

Patrice Newell was born in Adelaide. After careers in modelling, journalism and television, she moved to a farm near Gundy, New South Wales, with her partner Phillip Adams, where she produces biodynamic olive oil, garlic, honey, soap and beef. Her books include *Ten Thousand Acres: A Love Story*, a heartfelt call for sustainable land use; *The Olive Grove*, her bestselling account of leaving the city for life on the land; *The River*, a critically acclaimed examination of water-management issues facing rural communities; and *Tree to Table: Cooking with Australian Olive Oil*. You can follow the life of Patrice's farm at patricenewell.com.au.

Also by Patrice Newell

The Olive Grove
The River
Ten Thousand Acres: A Love Story
Tree to Table: Cooking with Australian Olive Oil

Who's Minding the Farm?

PATRICE NEWELL

VIKING
an imprint of
PENGUIN BOOKS

VIKING

UK | USA | Canada | Australia
India | New Zealand | South Africa | China

Viking is part of the Penguin Random House group of companies
whose addresses can be found at global.penguinrandomhouse.com.

First published by Penguin Random House Australia Pty Ltd 2019

Cover design by Alex Ross © Penguin Random House Australia Pty Ltd
Cover image Shutterstock/Scott Book
Internal illustrations Shutterstock
Typeset in 12.5/18 pt Adobe Garamond by Midland Typesetters, Australia
Printed and bound in Australia by Griffin Press, part of Ovato, an accredited
ISO AS/NZ 14001 Environmental Management Systems printer.

A catalogue record for this
book is available from the
National Library of Australia

ISBN 978 0 14378 939 0

penguin.com.au

For

Mary Newham nee Owens
23.12.1930–15.6.1987

and

Aurora Adams

CONTENTS

INTRODUCTION

WE CALL OUR PLANET EARTH, YET WE DON'T THINK MUCH ABOUT earth itself. As a proud dirt farmer, I reckon we should have a Minister for Soil. Not just a Minister for the Environment – we've plenty of those, at federal and state levels, and they rarely mention soil. When the UN formulated its seventeen Sustainable Development Goals in 2015, soil didn't rate a mention in those either. Yet soil is at the heart of achieving many of those goals, including the elimination of poverty and hunger; and clean water, sanitation and good health.

Soil: you'd think it was a dirty word.

While the UN was developing its goals, there were environmental crises all around the world, in this the Anthropocene period, a time when material wealth has finally come up against the Earth's biological capacity to support it.[1] The challenges and complexities are so overwhelming that it seems we're unable to make proper decisions. Do we believe that soil is as resilient as we are? While it's taken a beating, its natural ability to filter contamination has provided a buffer to total annihilation. But there's a limit. Some soils are exhausted.

Humans have always farmed within particular cultures, but with many traditions in common. Agricultural knowledge was once as basic as what our mothers taught us about washing our hands and cleaning our teeth. But unlike those domestic lessons, agricultural traditions are being lost. With ever fewer people working on farms, how could it be otherwise? Despite all the new technology, all the information now available in cyberspace, we've failed to pass on the simple things required to grow our own vegetables.

Inevitably it was Australia's pioneers who began the process of the great soil loss – the structural depletion, the acidification, the compaction and the nutrient loss, in a country whose thin soils didn't have a lot to begin with. Habits are hard to break, and those bad habits that started the destruction linger on, but the spread of drug-addicted industrial agriculture in this epoch of agricultural imperialism cannot continue. It must not be the future.

The fifth of December is World Soil Day.[2] It was officially adopted by the 68th UN General Assembly in 2013, but in Australia a much bigger fuss is made each year about a horse race in Melbourne and a yacht race arriving in Hobart. Soil is our lifeblood: it needs to have a fuss made about it. And since there's nowhere on the planet that hasn't been affected by the human species, we need to think not in terms of preserving some unspoilt ecosystem or a patch of pure soil, but in terms of repairing the damage. It's time to face the facts – and to try to change those facts.

I like to think of our farm, Elmswood, in the Hunter Valley, as an old/modern, small family operation, but it's still part of a global trade system. And no matter how we strive to do things differently, in a healthier way, there's no escaping the fact that we're connected to the farms next door, where our beliefs and practices may not be shared. And those farms are connected to the farms next to them, and to a world of agriculture in crisis. No farmer farms alone. We turn to each other in times of trouble, during drought or when bushfires threaten, and we often turn away from each other when it comes to views on the use of chemicals or climate change. Agricultural chemicals are manufactured synthetic products including pesticides, herbicides, highly soluble fertilisers and are the foundation of modern industrial agriculture. But manufactured synthetic fertilisers aren't doing the job any more, and plants are resisting herbicides. The whole farming sector needs to do things differently. The International Federation of Agriculture Movements (IFOAM), describes organic agriculture as a 'production system that sustains the health of soils, ecosystems and people. It relies on ecological processes, biodiversity and cycles adapted to local conditions, rather than the use of inputs with adverse effects.'[3]

Farms can and should be places of healing, for both humans and soil. I believed that when we bought Elmswood in 1986 and

I believe it today, and in this book I discuss how it can be done.[4] We're one of the minority who are trying to farm with nature, rather than against it; to rebuild the land after two hundred years of neglect or abuse. And while it can be hard at times to remain hopeful, I take heart from the fact that organic production is on the rise. Elmswood, a biodynamic farm, is part of the 35 million hectares of Australia classified as Certified Organic Agriculture. Most of that total is made up of grazing land in Queensland, but it represents 3 per cent of total farm gate production. That's a retail value of around $2.4 billion.[5]

The national mythology has the heroic pioneer braving a strange new land to make a life and a living. A large part of that myth, the idea of *terra nullius*, allowed vast tracts to be turned into private property, which was divvied up between white men, sometimes with pseudo-legal paperwork, more often simply squatted on by the entrepreneurs of the day. Farms and districts began to specialise – here beef, there wool or grain or wine or cotton, even such surreal crops as rice. Each making different demands on soil, demands that cannot always be met, particularly in the case of cotton and water-hungry rice.

What we have today, scattered across Australia, are the remnants of those nineteenth-century start-ups. Everything from small family holdings to industrial factories, where the product is beef, pork, chickens or eggs. Not far from us, city folk have bought a few hectares, but not to live on. It's land for fun. Somewhere for the kids to ride horses or roar around on quad bikes. For many newcomers, in fact, life on a farm is more about 'lifestyle', and it's more common to have a pet horse than a vegetable garden. This was evident during the recent drought, when donated hay was fed to horses while the farmers were given food parcels.

Elmswood is somewhere in between these extremes – 4000 hectares of land ranging from river flats to narrow valleys and steep hills where sometimes snow falls. It's a nineteenth-century homestead, a collection of cottages and huts and a shearing shed. But it's first and foremost about the soil, about leaving the earth in a better condition than we found it, for the future. But for whose future? Our daughter's? Or some unknown future owner? The question of who will be our farmers, and how they farm, is one that affects all of us.

It seems like yesterday, but it was 2009: I'm at Mascot international airport while our daughter, Aurora, seventeen, checks in for her flight to the UK. Or perhaps more accurately her flight from Australia. She's doing what more and more young people are doing, heading overseas to plan a future. In Aurora's case it's to check out the universities that have accepted her application, providing her HSC results are good enough. Travelling alone, she will couch-surf in London, Birmingham, Manchester, Durham and Edinburgh.

Phillip and I are bereft as the queue moves inexorably forward. I start to cry. Not yet in wrenching sobs, but with tears that swell slowly and that I try to stifle. I should feel proud of her confidence – and I am – and I want to share her excitement, but the wound in my heart is so deep it's hard not to think just about myself. Aurora checks in her tiny amount of luggage, gets her boarding pass, and the three of us sit in a coffee shop until it's time to say a final farewell.

The mood of that day resonated for months. I found some distraction in thinking deeply about the farm. Over the years, I'd had many conversations with friends about the moment when it dawns on them that their children may not want to stay on the farm and carry on the work. Even though I've never said – not to myself or to my daughter – that we're building an asset at Elmswood to pass on, there has always been the hope that one day she'll come back to live here and raise her own family.

But in the ten years since that first departure, Aurora has spent four years at Edinburgh University, several years in a demanding job in Sydney, and then she moved to Europe. This time with her partner, Susannah. So any succession planning involving her feels further away than ever. The farm manages without her.

I'm told by friends that you should never focus on family succession plans while your offspring are under thirty. That age may trigger a rethink and children sometimes come home to the farm. It's been happening a little around here, as those who left realise they don't enjoy the corporate world their university degree led them into, or they simply failed to get the right job. Farms can in fact absorb a large workforce, as long as people don't desire a high income or the latest technology, and can make do with poor communication networks, fewer services, and poor health and education options.

Back in 1987, when I swapped the city for the farm, I also switched careers, leaving my television job as host of *The Today Show*. My partner, writer and radio host Phillip Adams, and I originally bought 1600 hectares, then another 1200 next door, then a few years later an adjacent 1200 hectares of wilderness. It would take some fifteen years before I felt I could say with any confidence that I knew enough about farming to run our organic property, and to live up to my vow to protect the land. No official training is required in order to become a farmer. If you're professionally nursing a human, you need a degree. If you're nursing a calf or a crop, you can do it without any qualification at all. I became a farmer with no prior experience and I learnt from everyone and everything; from numerous courses on soil and plants, and from all the people I've hired to work here. Every single day on the farm has been an education.

I didn't always know precisely what to do but I knew what I wouldn't do. I wouldn't turbo-charge crops and pastures with manufactured synthetic fertilisers. I wouldn't clear more land. I wouldn't draw more water from our long-suffering creeks and rivers. I wouldn't use poisons, nor would I adopt a bad and broken business model to bolster short-term profits. Although we needed mechanical tools, I wouldn't automatically try to replace human workers with machines. Phillip and I wanted people to work on our farm. To share it with us.

Elmswood began its agricultural life in 1835, when the first white occupants pushed aside the Wanaruah people. Although the land has always looked to me as noble as any landscape painting, and has provided economic and social sustenance to many families preceding us, it was, and is, silently suffering from invisible ecological damage. Actions once thought important to development were carried out across the hills and valleys. Trees were chopped down along the river and creeks, in the belief that this helped the waterflow. But it doesn't, and never did. Instead the clearing caused bank erosion and silting downstream. The unshaded earth now feels the full brunt of our increasingly harsh summers.

Our predecessors planted all sorts of fruit trees, vegetables and grains, and numerous grasses – 'improved pasture'. They brought in sheep, horses and cattle to eat those grasses, and slowly the vegetation began to change. The waterflow began to speed up and the gullies widened and deepened. Soil was blown away and the new pastures became dependent on fertilisers to keep them flourishing. Droughts came on average every decade.

These days we better understand global weather patterns and can predict droughts to some extent, but living on the land will never be easy. Many of my farming friends are so busy these days they don't milk a house cow, keep chickens or ducks, and they rarely kill and butcher animals for their own meat consumption. A lot more of their time is going towards paperwork and keeping up-to-date with regulations.

Over the generations, Elmswood expanded during the good times and shrank in the bad. Sheds and worker cottages were built, some of which are still standing; others are only a few stumps in a paddock. In 1880, one Laban Wiseman started building a homestead, a proud, two-storey brick statement on a knoll overlooking the junction of the Isis and Pages rivers. At that time the New South Wales Department of Lands Register recorded a hundred carved trees standing where the rivers met, proof of a major Wanaruah meeting ground. Sadly, no carved trees now remain. We're told the area was called Gunda Gunda by the Wanaruah people but now we call it Gundy, a small village next to our property.

The head of these two tributaries of the Hunter River is recognised today as the northern edge of the Hunter Valley. The site is a three-hour drive from the mouth of the Hunter River at Newcastle, the world's biggest coal port. Up past our place, over the Liverpool Range, the land opens out into a much bigger catchment – the north-eastern limit and the head of the great Murray–Darling Basin. Our Hunter Valley catchment is smaller and coastal.

Within this vast and complex geology, my commitment to the task of growing healthy food across 4000 hectares is frequently confronted by failure, and around me is evidence of the defeat of others. Fewer farmhouses are being renovated in the district, vehicles are looking more beaten up, and farmers are getting older, including me. Some struggling farmers continue to make the same mistakes of old, overgrazing and destroying native pasture – the start of a trajectory that often ends with selling off paddocks to make ends meet. The produce from the farms in the

Hunter has also changed over the years. Like Elmswood, they were once totally dependent on wool and beef; but they've been branching out and now breeding horses of various kinds is big business, medium-sized dairies operate, grapes are grown for wine and small- to medium-sized olive groves have been established.

Passing empty paddocks on the way into town can make it seem that the rural sector is in retreat, and on road trips across the country it's often hard to comprehend where exactly our $61 billion agricultural sector is.

The juggernaut of globalisation has been inescapable for Australian agriculture. From the start, merino wool was a major export earner, and it still is. Australia has 70 million sheep and wool reaches markets far and wide. Overall, 70 per cent of all our primary produce is sent overseas, including much of our organic produce. Agriculture is as much an export business as mining, so when I see asparagus from Peru, garlic from Mexico and cherries from California on supermarket shelves, I wonder why we still need to import so much food. The answer is simple: all-year-round supply, and money. Cheap is the name of the game. Cheap food is as political as cheap power, and the supermarkets are happy to comply. The global shopping cart is at a store near you and on your computer.

But in exporting agricultural produce we are also exporting our soil and water. Every time we sell something that's been born, shorn, cut or harvested – from cotton fibre to goat meat, from wine to my beloved garlic bulbs – a little bit of soil has changed, often sacrificed, along with as much as 70 per cent of the nation's water supply.

Far from a feeling of professionalism and prosperity, Australian farming's arse is hanging out of its trousers. Where are the neat and tidy sheds housing the shiny machinery? Not on our place.

We regularly hear that the world will struggle to feed its people. I don't agree. That is to say, I think this situation is inevitable only if we keep proceeding the way we have been. In Australia so many fertile paddocks lie idle – land that is capable of storing a lot more nutrients and water and therefore producing more food. And this is possible without adding billions in inputs. The problem is a lack of people who know how to manage soil properly, and a lack of people wanting to do the work. At the core of it all is a poor understanding of how to harmonise with our unique, varied and complex environments.

While Elmswood now has a variety of agricultural operations, the land itself is still officially designated into classifications. These exist for the statistical measurements carried out in surveys and the agricultural census. Each year, I fill out a form and send it off to a government office, declaring the number of commercial trees on the property, the variety of crops, the extent of the native vegetation, and an estimate of the number of domesticated animals. Most of our land is classified as being under 'broad acre management with beef and sheep', though the river flats are devoted to horticulture, with garlic production, and six thousand olive trees in an established grove. And all across the property there's apiculture, or beekeeping.

Our farm isn't a perfect site for olive trees. We learnt the hard way, first killing some with kindness, planting others in frost pockets. While the layout of the grove was designed for light and access to irrigation, some sections have poor soil, despite our best attempts to improve it, and bad drainage. And with summer heat now threatening to melt the thermometer, it's uncertain if either olives or garlic will be viable in the future. In the here and now, I'm constantly rethinking the things we produce, the way we do things, and the way we sell things.

For the three years of 2016–18, the summer crops on the farm have failed. They germinated, starting their life under a scorching sun, but could not be irrigated due to drought, relying instead on the existing moisture in the soil. Ten straight days of temperatures over 40 degrees will wilt many plants. The absence of the lush rich green cover crops we once produced meant that the soil's organic matter was reduced, and the structure of the soil became less friable. We've pruned the olives into smaller trees because we can't guarantee them water, and a smaller tree needs less water. But as we've pruned – with the help of the sheep, which nip the lower branches – the tree trunks have been more exposed and, as a result, get sunburnt. If temperatures continue to climb we'll need to change the pruning shape to better protect the trees.

Meanwhile the fire permit season has been extended: in 2018 it started in August. And it looks set to become a three-season affair: spring, summer, autumn. This means any small fire needs a special permit from the local fire brigade otherwise no fires are permitted. When a welding accident close by sparked a bushfire, we banned outside welding in summer and discouraged smoking. If this is the future, the seasonal calendar will have to undergo many more adjustments, with winter becoming the time when many tasks can be completed.

And yet the conversation about climate change has been alarmingly slow to progress. Denialism is entrenched. The coal-mines of the Hunter Valley bring in local money and provide state funds via royalties. They are also a source of greenhouse gas, so it's easy to understand why there's resistance to change, but the conversation is little better elsewhere in Australia. Farming and land clearing have also driven climate change. The agricultural sector in Australia is the third biggest emitter of greenhouse gas, after electricity generation and transport.[6] A tractor pulling a plough releases carbon instantaneously with every pass. Clearing plants that we dismiss as 'scrub' also pushes carbon dioxide into the atmosphere. So does spreading nitrogen-based fertiliser. But there is a way to redress this, and that's photosynthesis, the miraculous process that plants perform naturally, removing carbon dioxide from the air and earthing it in the soil via their roots, also releasing oxygen along the way. Plants not only hold the soil together, they help hold soil carbon.

Photosynthesis could heal the world if we let it. Every farm can be part of a system to manage the climate. One of the worst sights during this latest drought has been the slow denuding of hillsides, the turning of pasture to dust. Thousands of bare hectares without plants is not using the sun to build the soil and sequester carbon.

Agriculture needs to be driven by environmentalists. That statement, I know, scares a lot of people. At a conference in 2008 of around fifty farmers who identified as sustainable, I asked people to raise their hands if they saw themselves as environmentalists. A grand total of five hands went up, and they were tentative and apologetic. Wavering wavers. I'd asked my question assuming that the whole audience would be proud wavers (I should have borne in mind the saying of barristers: when cross-examining, never ask a question for which you don't already know the answer). Even among progressive farmers the word 'environmentalist' is problematic and divisive.

Yet environmental policy must begin at the farm gate, because it's farmers who own the land, the soil – or at least that land that isn't mortgaged to the banks – and they're the people who need to undertake the repair work. A big reason that many farmers won't put their hand up comes down to a collective loathing of 'environmentalists', who, to be fair, have yet to learn how to talk to farmers. We have a classic failure to communicate.

Environmental damage began in Australia with the First Fleeters of 1788, who were met with the gift of clean water, trees galore for timber, and beckoning grasslands. What could possibly go wrong? Tragically almost everything, from day one. Australia is a country of vastly different soil and land types within short distances of each other. The settlers here did not let the land itself define its use. Different fibre and food crops demand nutrients and water in different ratios; different animals require different terrain. We might want to increase production or diversity but that doesn't always mean we can, at least not without first carrying out some restoration. Respecting the capacity of the land, of all our different landscapes, is essential if we are to avoid further disaster.

CHAPTER I

THE GARLIC PATCH

I'VE ALWAYS GROWN GARLIC IN THE VEGETABLE GARDEN AT Elmswood, having long been fascinated by it as both food and medicine. There's so much more to this amazing relative of the onion than flavour and aroma, symbolism and superstition. With almost three thousand scientific papers evaluating its medical properties, it's fair to say that the health benefits are many, from its basic antimicrobial properties to protective effects against serious health problems. Yet it wasn't until 2007 that I planted it as a marketable crop.

Graeme, a retired vegie grower, taught me how. I met him one autumn at the annual meeting of the Biodynamic Association of Australia, held that year in Merrigum, in Victoria's Goulburn Valley. He looked a bit like Chips Rafferty and he was at the meeting because his daughter and son-in-law were growing biodynamic apples. He'd be passing by Elmswood soon, he told me, and asked if he could call in. It was more than a year later before he turned up.

When he did he brought with him a bulb of garlic. It was purple, 7 centimetres across, with good-sized cloves. A few years earlier Graeme had been walking down a street in Kyabram, near Merrigum, and came upon a bloke gardening in his front yard. Graeme stopped to admire his garlic, then talked the man into giving him some bulbs. Over the next few years, he turned those thirteen bulbs into enough for a modest commercial crop, using a small patch in a friend's apple orchard.

Graeme and I found we had much in common. We both rose early and ended the day early. We liked long road trips across Australia. He liked growing vegetables, I liked cooking them and we both liked eating them. His wife had died a few years earlier from breast cancer and now, at the age of sixty-three, he had no fixed address. But he didn't feel homeless, just free. He always travelled with a swag, had no interest in a bed, and needed to be as close to the earth as possible at all times.

When he returned to Elmswood a few months later I told him his garlic bulb had been delicious, very sweet and juicy. He was horrified. He'd expected me to plant the cloves, not eat them, and had come to inspect my newly sprouted plants.

Graeme urged me to grow garlic on the river flats in front of the homestead. Once our pride and joy for producing bales of fine lucerne hay, this land had been reduced by the drought of 2001–07 (referred to now as the Millennium Drought) to just another paddock for stock rotation. I eventually agreed to a crop, on the condition that Graeme helped me. He promised to call in periodically on his trips across the country.

He arrived at planting time with boxes of purple garlic gathered from his friends and family's test plots as seed stock. These days we don't plant small cloves, but back then, with seed stock hard to get, every bulb was carefully opened and each clove carefully separated. Garlic does not produce true seeds, but grows from the clove. This means that a little extra must be grown each year to ensure the next season's planting material.

After deep-ripping the soil and using the tyre tracks to measure out twenty 50-metre rows, we dropped down on our hands and knees and gently pressed the cloves into the soil, about 3 centimetres deep and leaving 10 centimetres between each one. (Organic farmers don't plant as close together as farmers using synthetic manufactured chemicals because they rely on soil biology to feed the plant.) With five hundred cloves a row, and twenty rows, we'd have ten thousand bulbs for the first crop.

We hooked up the overhead irrigation pipes we'd used for the lucerne crops and turned them on for the first deep watering. Seeing the rows of strong green garlic shoots peeking from the ground two days later was cause for much excitement, and in the following weeks I checked the crop during my morning walk. I had no idea if the garlic was thriving or stunted, but it always needed weeding – that much I knew.

The relentless work of weeding marks a clear difference between organic and industrial farming. We do it the slow way, the hard way. And weeds are more of a problem with a garlic crop than with other vegetables, because of the longer gestation period. Garlic is in the earth for between seven and nine months, whereas lettuces, for example, are sown and reaped in as few as six weeks. But if chemical warfare is the price of an easier life, it's a price I refuse to pay. At least we work without needing protective clothing, knowing we're not slowly killing ourselves or the earth with poison. Organic farmers want to grow things in good soil, in living, breathing soil. We don't want to mount indiscriminate killing on the microbes within it, or on the plants, the bees, the birds. Our mindset is different.

During that first year, the garlic patch was a classroom where Graeme gave me lessons. There was the lecture on soil moisture. After irrigating for an hour I'd turned off the pump, confident that the garlic was well and truly watered. Graeme knew otherwise. Instructing me to kneel beside him, he dug his hand into the moist soil and pulled out a sod. 'You've short-changed it,' he said. 'Vegetables are quick-growing plants, so they're heavy drinkers. And this soil isn't anything like wet enough. If it was, it'd stick together.' He opened his hand and the soil fell apart. Too dry. I got it.

When I look back, that initial year was pretty easy; I didn't know enough to be worried. With Graeme coming and going, I shared most of the work with Colin, a young bloke from Scone whose main jobs were in the garden and olive grove, but when he had spare time he'd wander over to the garlic patch and pull some weeds. For heavier work I'd call on Jamie, our new farm manager, who'd leave stock work or the repair of sheds and fences for a few hours. He wasn't enamoured of tractor work, though he'd sow and harvest. Somehow, our workforce of just four part-timers, Graeme, Colin, Jamie and me, managed the tasks.

Seven months after planting, we harvested the garlic, again by hand, carefully lifting the soil around the bulbs with garden forks. I sold part of that test crop to wholesalers, who said the quality was fine – proof that we could grow garlic at Elmswood that met the market standard. Most of the garlic from that year we kept for seed stock. Thus began our garlic business, and Graeme became part of our lives.

I saw potential in purple garlic, which was not common at that time. The second crop was far bigger: thirty rows of 100 metres each, with a thousand cloves per row, amounting to thirty thousand bulbs. Most of that was purple garlic. When you're deciding which variety of garlic to plant, you need to consider what will grow well in your climate, in your soil. You also need to consider whether the cloves break easily from the bulb, whether the skin peels off easily, whether it will be the right size to fit nicely into a garlic crusher, and, most importantly, how it will taste.

We're always testing new varieties, and also trialling different row lengths and widths, trying drip lines on top of the rows, then drip tape buried between rows. Sometimes we mulch thickly, then we mulch lightly; we use woodchips one year, hay another. No two crops have ever been grown the same way. There's an old belief amongst garlic growers that you should grow garlic in fresh soil each year. For the first six years we followed that approach, but we've now returned to that original patch and sown it with garlic again. These days, garlic growers are divided on the issue of needing fresh soil every year. Healthy soil is the issue. The soil determines the quality of the garlic, and the soil in every patch will vary. A heavier soil will usually produce bigger bulbs than sandy soil, but the bulbs emerge cleaner from sandy soil – you can often almost dust the soil off those bulbs.

There's no sure-fire calculation for the relationship of the area planted to the magnitude of the final crop – it's all guesstimates. And because even bulbs from the same soil are not all the same size, we're never sure about volume until we've set aside our seed stock and discarded the damaged bulbs. The industry target is 320,000 bulbs of garlic per hectare, but I prefer to start with the tonnage I'd like to sell (ie. 10 tonnes), add the volume needed for seed stock (800–1000 kg), and factor in 10 per cent wastage. Although nothing is really wasted – leftover garlic is given to family and friends, with the detritus going to the sheep.

By the time we'd planted the fourth crop, I was urging Graeme to move to the cottage. Hesitant at first, he slowly started staying more often and for longer, though constantly reminding me that he was not, absolutely not, settling in. He was most emphatically not taking a full-time job. He was a retired vegetable grower obsessed with garlic and was just helping out. He must be free to come and go. Whenever he was at the farm he and I talked nonstop about garlic. Every year there were hard lessons. Sometimes the weather was good, other times it decimated the crop. A good year saw a yield of 7 tonnes from one hectare.

Most garlic varieties are planted in autumn, although some can be planted in winter, depending on the climate zone. Any region with cold winters and hot dry summers should suit a garlic crop, and it can grow in any good, well-drained soil as long as it's well watered and weeds are kept under control. Here at Elmswood we plan each crop two years ahead so we can prepare the soil.

Our preparation is based on growing what's commonly called a 'green manure', which is ploughed back into the soil when mature. Lots of crops can be used for green manuring. We like oats and mustard in winter, cowpeas, lablab (a type of bean) and millet in summer. They all provide organic – carbon-based – matter, and help form a light, fluffy textured soil that's ideal to plant in.

Once the cloves are in the ground and have been deep-watered they shoot up very quickly, to grab the last of the summer–autumn solar energy before resting during winter. By this stage we want to see a deep, strong root system. Winter is a time of wait and see, and to be alert for things like aphid attacks. Usually there's a point at which the weeds do get away and the team and I look at the patch in despair. One year the weeds won and we had to surrender a section.

I reflect on this when reading about the many chemical options available for the farmer. Everything in organic and bio-dynamic agriculture is about building the soil so that it can be productive without using poisons, and there are alternatives with which to control pests. Every spring, when the nights warm up and it's getting close to garlic harvest, moths and other insects arrive in the millions. Some mornings we'll find immense numbers of dead insects from the night before, but sometimes not all the visitors die. When a thrip was spotted on the garlic one year, I had a slight panic. These minute flying bugs are elusive and like to hang out during the night at the base of a plant.

To deal with them, I melted some of my olive-oil soap trimmings in hot water to make a sudsy solution and poured it into backpacks, to be sprayed along a test row. I'd used this solution before, on the olive trees, spraying the leaves to help suffocate olive lace bugs. I was applying the same principle to the garlic, hoping to suffocate the bugs.

The garlic was drenched and I couldn't see any thrips flying around. They were either not there or dead. What if they were hiding deep within the garlic leaves, having survived our efforts to suffocate them? If they lived on in numbers there'd be a risk that they'd nibble at the fresh garlic. There'd be holes.

Days later I couldn't see any visible change, so we did nothing more. Bugs come and go, I reminded myself. We often find that insects like lace bugs and thrips appear and disappear like a sore throat. Their arrival triggers the fear that they could be the start of something serious, but more often than not they fade into oblivion.

My witch's potions are not always successful. Take my anointing of the bush lemon tree outside the kitchen. Convinced I'd be doing it a favour, I once sprayed it all over with a brew of garlic solution: a single bulb crushed and soaked in water, then strained for spraying. Months later, when the lemons looked magnificent in their full yellow glory, I plucked a dozen to make one of my signature dishes, a ricotta cheesecake. I carefully zested all the rinds, only to discover when the cake was cooked and served that it was not a lemon cheesecake, but a garlic one. It did not meet with the usual appreciation.

The overwhelming challenge for Australian garlic growers is the cheap imported product, which is guaranteed to have been sprayed with methyl bromide in order to qualify for a grocery shelf: this is a legal requirement. This chemical, used to kill mites, has been associated with prostate cancer. Back in 1987 the UN decided that methyl bromide should be banned because it was depleting the ozone layer. The Montreal Protocol called it an 'extremely' dangerous vapour'. So while it's banned almost everywhere in the world, and Australian farmers aren't using it in their silos anymore to kill insects, it's still used in Australia for biosecurity purposes on all imports of fresh food.

If you want your food fumigated, cheap imports are the way to go. If you want clean food, and also clean water and soil, with Australians doing the work, local is the way the go. But as of 2018, there are still only a hundred or so commercial Australian garlic producers, while China boasts around 2 million.

In China, garlic is usually grown on mixed farms, along with hundreds of other small crops of less than a tonne. The garlic you'll find in a Chinese market or domestic kitchen is much more likely to be smaller, and not the white, round, bleached-looking product that is globally traded. According to the UN's Food and

Agriculture Organization, in 2016 China produced 21.2 million tonnes of garlic, around 80 per cent of the world's market. And China's garlic isn't just found in the netting bags you see in supermarkets; it's a foundation ingredient for the industrial food sector. Where would bottled pasta sauce be without this important, all-purpose flavouring?

But production in China may have already peaked, as new growers in Ethiopia, Brazil and Bangladesh enter the market. Most importers in Australia get supplies from established growers in Mexico, Argentina and Spain when China's cheap stocks run low.

So while I'm always taste-testing different garlics, I avoid the imported bulbs because of their chemical drenching.

Garlic belongs to the genus *Allium*, as do onions, leeks and chives. Each of these plants contains varying degrees of sulfur compounds, which provide the odour, the taste and sometimes the tears. These same compounds give garlic its legendary medicinal properties. What packs the punch is the allicin, which is produced when a clove is crushed or chopped: an enzyme called alliinase converts an odourless compound called alliin into the allicin. This same process also produces garlic's powerful aroma. So while the bulb sitting on your kitchen bench is a storehouse of nutrition and medicine, it is only activated when you crush, smash or chop it up.

I take comfort in knowing that when adding a fresh garlic

clove to food I'm continuing a thousand-year-old tradition. And before a garlic crop is ready for sale I'll be eating a bulb a day plucked straight from the paddock: that's for both pleasure and health. Our styles might be changing but our bodies' physical needs are not so different to those living in Egypt five thousand years ago. While all true garlics have medicinal compounds, they're unlikely to be standard across all bulbs, just as every fresh ingredient will vary in its vitamin and mineral components. The active chemical constituents, trace elements, enzymes and amino acids all vary according to varieties and the way they're grown. Goodness can be found in dried garlic too, but again this varies. Every time I read a scientific paper on the qualities of fresh garlic it confirms my view that food is our best medicine.[7]

Nor are all garlics equal on the palate; some have a harsh flavour, some mild. Some have sticky skins, some have big bulbs with small cloves, others are small bulbs with big cloves. Our white garlic has been hard to sell because customers associate it with the white Chinese imports. I've failed to convince them of its virtues, in person and online. I love its pretty pink hue, and how it gets hotter in flavour as the season progresses.

We used to offer a late-harvest white softneck, affectionately called Ugly Garlic by the packing team. (There are two main sub-species of garlics: *orphioscorodon*, the hardnecks, and *sativum*, the softnecks.) It looks untidy, with thin skins that flake off easily, and is a magnet for mould on a humid day. It has a very strong flavour when fresh, lasts for 10 months and turns honey-sweet when baked. But when placed alongside the purple, all eyes end up on the purple. It was too hard persuading people to try the Ugly which I now only grow in my kitchen garden.

When I sit around with fellow garlic growers and taste-test different varieties, we know that if we were to do the same test in a month's time, the flavour and our assessment of it will change. Having peeled off the outer skins and opened the bulb, we count the cloves and check them for firmness, colour and any disease. We have a sniff, making sure there's no mouldy aroma. Next we slice the clove thinly, wait at least a minute for the aroma to develop, and then have another sniff. This should reveal the fresh pungency of the garlic juices. Finally, we have a taste, biting into just a tiny slither. This is the best part, but after four or five samples it is also the most tricky. The nutty freshness of garlic can also have a strong heat that fills the whole mouth. The residual hotness of the previous garlic can be hard to eliminate, and the volatile oils are sometimes slow to develop in the mouth; as a result, the tastebuds get confused. We always have parsley and sliced apples on hand to cleanse the palate.

Personal preference for the heat of a fresh clove can vary as much as tolerance to chilli. When I judged flavoured table olives at an official show, it intrigued me just how different chilli can taste. That's because there are so many different types of chilli, and ways to present it – thick flakes, dried to a powder, sliced fresh. A chilli left too long has a tired, stale aftertaste; the same goes for garlic.

Garlic is now officially judged in many national and local shows, with most Australian competitions occurring in February, when our Elmswood garlic is past its prime and I'm already thinking about a new crop. By February I've been eating our fresh garlic since the previous September, enjoying its changing flavour profiles. For commercial growers like me, having eight perfect bulbs to send off to be judged isn't the point. I need tonnes of it to turn out well. While contests can be fun and help generate garlic appreciation, growing an award-winning bulb doesn't guarantee that the rest of the crop is of the same quality.

There's no end to the development of good garlic, as growers continue to experiment with varieties and growing techniques. Around the world, the appearance of garlic varies with the soil type, geographic position, altitude, moisture, weather and the way that it's grown.[8] If you were to take one bulb and plant each of its cloves in a different place, chances are that each new bulb would look and taste a little different. And each year's crop will differ from the previous year's, even when planted in the same area. Complexity makes garlic the most fascinating and frustrating of crops.

Because of these variations, there can be confusion in identification. A grower at Williamtown, near Newcastle on the central coast of New South Wales, came to Elmswood with the garlic he'd grown from my purple Glamour seed stock. His garlic was white, and the bulb formation was different from anything I'd seen, before or since. I couldn't believe it was from the same genetic stock.

I still test new varieties, but not dozens per year as I once did. In the beginning I wanted to grow as many varieties as possible, especially for the flavour variations, but after a few years it became too complicated and difficult to keep detailed assessments on so many types. Trying to harvest each variety at its peak is hard enough, without some bulbs requiring four extra weeks to mature. Harvest is such an intense time; we're like storm chasers, and if it looks like rain I panic that the entire crop will be wiped out. And sometimes those late-maturing bulbs were planted in the middle of a patch that had already been harvested. I can't know when I plant a new variety whether or not it will take longer to reach maturity.

Australia's nascent garlic industry has no official garlic register yet, whereas the US, for example, has a system that registers the many varieties available. Because garlic is both food and medicine, knowing which varieties have the finest chemical components will become important as our industry expands. The quality of food and people's expectations of it are subtly adjusting all the time. (I'm reminded of this whenever I buy a watermelon and hardly find a seed: how different to my childhood memories.) The USA, as well as China, Japan and Europe, also funds research on garlic's nutritional and medicinal value. In South Australia tests using aged garlic were shown to lower blood pressure.[9]

Along with learning how to grow garlic, I also needed to learn how to sell it. Small producers seek to find a niche with high-quality products, or to find ways to add value, as I do by making soap from our olive oil and honey. The handlers – the people who transport, store, wholesale, retail and import products – are powerful, even in organic agriculture. So too are the processors – those who convert raw produce into goods like wine, baby food and ice-cream.

Before selling a crop, I need to quantify it, so it's back to guesstimating the gross weight. Guessing is an important part of a farmer's life. How much wool will the shearers shear? How much wheat will be harvested per hectare? How many kilos of olives per tree? When running the cattle up the race we guess at their weight before they reach the scales. Guesses become wishes.

You'd think guessing with garlic should be easy. We plant precisely, with cloves spaced evenly apart. Simple multiplication, surely. But how many bulbs will be too small to sell? How many will be damaged? Too bruised, broken, split, water-stained or misshapen to make the final cut? How many will be plucked out of the ground by cockatoos or kangaroos?

As the bulbs grow, we scratch around to see if they're filling out and try to predict their full-grown size, then make a quick calculation of gross tonnage. But garlic is a fickle crop, easily damaged, and requires up to four weeks' curing after harvesting, during which time much can go wrong.

Producing olive oil is easier. It has a longer shelf life, whereas garlic, being a fresh product, demands a more immediate delivery system. Selling a few hundred kilograms to a wholesaler had been straightforward enough, but we needed to secure our own market. Not that I'm against wholesalers; they do a good job distributing products around the country. But if you can sell direct, why wouldn't you? It's one way around the commodification of supply and the stranglehold of the big supermarkets, who prefer to keep suppliers to a minimum – a system that inevitably results in diminishing diversity.

Selling direct was where Roger came in, a graduate in agricultural economics from Sydney University who lives in the northern beaches and ran an online home delivery service for organic produce, including my olive oil. He's a passionate advocate of good food. His Polish mother had taught him well. She and her husband had escaped the Holocaust that killed many of their relatives, raised a family in Australia, and professionally thrived in academia and business.

Fond of wearing a trilby and loud floral shirts, Roger looks like a muso searching for a jazz band, but he's a true crusader for the family farm. I remember his long pause when I phoned to ask about adding garlic to his marketing menu. He told me he'd call back the next morning, and when he did he said, 'We'll put the garlic in little boxes. I can get the boxes made.'

'But where would we do the packing?'

'Don't you have a shed?'

We had lots of sheds, but none suitable. Still, I decided to have a hay shed modified and give it a go.

Luckily, manager Jamie was an accomplished builder. He strengthened the hay shed with steel pylons, bolted Zincalume onto three walls, and poured a concrete floor. Voilà: a garlic shed. And Roger's boxes duly arrived – purple, to match my favourite species.

Next came the preparation of the crop for sale. We brought the bunches of garlic in from the shearing shed, where they'd been hung to cure in the breeze, and spread them on tables once used for wool sorting. These have slats so that all the dirt and skins fall through to the floor, keeping the table surface clean for the garlic. During preparation, every single bulb of garlic is individually handled. The first stage is called polishing, although we don't literally polish, and involves removing the outer skin and any residual soil or damage and then clipping the stem. We clip each variety in a unique way, some with long stems, some with none. I prefer the stems left on when gathering a clump for the kitchen. I tie them with string and hang them somewhere handy. Our purple Turban variety is trimmed to leave a 2-centimetre stem that somehow feels right. Some stems are dry and stiff by the time the garlic is fully cured, and small hands can find it hard to cut them. Sore hands and blisters are inevitable after a day spent polishing.

Next the garlic is placed in crates and taken to our little assembly line, where the boxes are filled. To prevent the bulbs being bruised in the post, we pack them in woodwool – the shavings of fresh pine from the timber mill – like eggs in a nest. The woodwool is biodegradable and any leftovers are donated to the chooks for soft bedding. Once filled, the boxes are stacked up at the far end of the shed, ready to be placed in Australia Post containers for their journey to a sorting division, before being despatched to customers.

As each year's garlic crop increased, we had to find more curing space. When we ran out of room in the shearing shed and hay shed, we made modifications to other sheds, always ensuring maximum ventilation. Garlic is full of water when harvested and each bulb needs dry air around it. If the weather doesn't deliver this we help create it with fans. One garlic grower even built an expensive dehumidifier.

More crates were also needed. Roger found a nursery on eBay that was selling thousands that had once been used for the floral bulb industry. Climbing into the truck, I headed for Peats Ridge, about two hours away in the Central Coast region. Peats Ridge was developed long ago by small orchardists. The old nursery was now a demolition site, and in fact the entire hillside was being bulldozed. Only one shed remained. Thousands of crates were stacked along the wall; they were perfect for my garlic. The manager told me that a new, state-of-the-art glasshouse was being built to grow cucumbers for the supermarkets. It would be temperature-controlled using biomass. Bioenergy: at least that was good news.

Loading what we needed onto the truck, I headed back down Peats Ridge, stopping at a run-down vegetable stand with an honesty box. The produce looked sad and defeated but out of rural solidarity I bought some anyway.

The following morning I told the story of the enormous new greenhouse, and one of our contract workers shook his head in disapproval. He told me how good it was to be working outside rather than in a glasshouse, where the chemicals had made him feel sick.

Food production shouldn't make anyone sick. One of the benefits of working with garlic is that you're handling a natural antibiotic. We sell cured bulbs (that's where the maximum nutrition is) but garlic can be eaten as a fresh plant any time it's growing, even as a substitute for spring onions. The stalks and the flower heads can also be eaten.

Our growing business also demanded more people to pack the garlic in boxes, and the question became how many could be crowded into the garlic shed without falling over each other. And where could we find them precisely when we needed them? Roger had an idea.

His friend Jimmy, a dentist from India, was doing a master's in business at an Australian university and was already part of the casualised economy at Australia Post, where he worked as a contractor delivering parcels. He agreed to come bush for a weekend and help, and to bring his wife and some friends. That's how we came to have twelve Indians and one Pakistani packing the bulk of that year's crop, working to Bollywood soundtracks booming from a speaker box hooked up to their iPhones. Multiculturalism was celebrated – many of our local team had never met an Indian, a Hindu or a Muslim. And there was the bonus of learning how to make their potato and pea curry, using loads of fresh garlic.

With garlic becoming as important to the farm's income as beef and olive oil, our next hurdle, which proved a much greater obstacle than refitting sheds, was the speed of Elmswood's internet. It wasn't merely slow, it was often stationary. We started business with dial-up speed. It helped that Roger had faster internet in Sydney and that we both had IT-savvy daughters living at home. Roger was able to back me up when things went wrong. Which was always.

Our shoddy internet isn't just an issue for selling garlic. It's also a safety issue – we need decent communications for the fire brigade during bushfire season, for health emergencies, for road accidents. Over the years, we've tried everything at Elmswood, wasting thousands of dollars on different satellite dishes, wi-fi routers and mobile devices. Ten years on, we're still waiting for the NBN and my iPhone often has a one-bar signal. Sometimes the only way to make a call is to drive up a nearby hill. Netflix? Forget it. The simplest downloads take forever. We're not alone with the problem, and a group called the Regional, Rural and Remote Communications Coalition has been formed to lobby for better internet access. Everywhere I go in remote areas, farmers are relying on rusty landlines and slow satellites.

In 2018 Phillip moved his Sydney home to a flat in Potts Point, and the first thing we did was connect to the NBN. The flat is small, yet suddenly the outside world was available online and it was our farm that now seemed claustrophobic.

After decades of slow motion we were travelling at the speed of light. But my delight intensified the rage at what we were being denied at Elmswood. No wonder people were buying everything online. It was so easy. If you lived in Sydney.

Alarmingly, our early garlic sales were as slow as the internet. But starting slowly turned out to be a blessing, allowing us to put better systems in place. Between Roger and me, we got around most problems. My largest fear was, and remains, that Australia Post, in its monumental indifference to Australian farms, would inadvertently destroy us while rushing to aid the business plans of international giants like Amazon. Australia Post is quite capable of servicing our needs but we've found its focus is wrongheaded. It was investing hundreds of millions of dollars in systems to more efficiently distribute goods bought from overseas, while the development of systems to get Australian goods from rural areas to the cities, towns or overseas felt lacking. Posting between capital cities is easy and relatively cost-effective; the price hike starts when your parcel leaves a small town like Gundy and heads to an equally small town in, for instance, northern Queensland.

Not long after we started selling online, Australia Post set up something called Farmhouse Direct, as a way for very small producers to get into the market, and we were convinced to try it. I thought it might provide a simpler alternative to what we were doing with our little online shop, which slowly but surely was attracting loyal supporters. But whoever designed their system had obviously never worked on a farm and didn't comprehend a slow internet or the cost of transport from isolated places.

And transport costs are disproportionate with very small orders. We all need quick, reliable food deliveries, and if this can't be done cost-effectively, a business can't survive.

Worse, Australia Post's system required us to forfeit the right to call the customer our own. Whenever someone orders through Farmhouse Direct, Australia Post gets the consumer's shopping history for their database. The system is set up for them, not the seller, and removes the ability of the farmer to communicate directly with the customer. This is anathema to what we try to do at Elmswood, which is to develop a personal relationship with each customer. We like to know who they are and we adhere to the view that buying something on the internet should be as personal and pleasurable as possible.

We endured glowing media accolades for Australia Post's now departed CEO, who, powered by his astronomical salary – far and away the highest in the entire public service – was supposedly leading the business into the next century. I eventually learnt from retired Australia Post workers what we'd known for years – that the inside story was chaotic. Whistleblowers described scenes from an updated version of Chaplin's *Modern Times*. There were new-fangled, turbo-charged conveyor belts that, in an attempt to speed things up, went too fast and threw the parcels off, particularly when going round corners. Confident of success, Australia Post had sacked a lot of people, whom they were forced to rehire. They needed staff to pick up the parcels when they fell off the belts.

The demands of Australia Post on our business were crushing. Central to the new hardware were barcodes: every parcel, big and small, had to have this electronic fingerprint. We were required to place a little sticker by hand on each parcel, proclaiming it to be untracked – in my view, an open invitation to theft. Why not just omit the sticker? Because the computers would not allow a parcel onto the fast-moving conveyor belts without one – it would be thrown off. This time deliberately.

Then came the pressure to move to something Australia Post call an eParcel, which meant we had to coordinate their barcode system with our shopping cart, which meant redesigning our perfectly good, customer-friendly address-label-producing system. We caved in and did it, paying the additional cost per parcel ourselves. Forced to fit in with their system at the expense of ours. We were told this new way guarantees delivery. It doesn't. Moreover, to find a lost parcel within the system is like needles in haystacks, and is even more annoying when someone orders garlic and finds it's been delivered and eaten by a household a few doors down the road. Customers who complain to Australia Post when this happens receive emails abrogating all responsibility. The parcel *has* been delivered, they are told, and the fact that it was to the wrong address doesn't seem to count.

Another obstacle we faced as sales started to pick up was anachronistic state laws. Over the years, the customer has become more discerning. The original question was: Where can I get Australian garlic? Now it's: What different kinds of Australian garlic can I get? More and more good garlic is being grown and sent around the nation – but only as far as the incomprehensible nineteenth-century laws allow. Here in New South Wales, we can send garlic to the Northern Territory, Queensland, the ACT and Victoria, and as of 2019 South Australia but not to Western Australia, or Tasmania. Those states are off limits, yet paradoxically they can send anything to us.

Investigating why these two states had quarantine restrictions, I learnt that it wasn't due to biosecurity. I could in fact sell the entire crop to those states if I paid for certificates – one on every single box – to prove the garlic was disease-free. (If I sold 5 tonnes to one person, one certificate was enough, but if I sold 5000 1kg boxes to five thousand people, I needed 5000 certificates). Quarantine restriction is really protectionism. If free trade is a global ideal, shouldn't it also apply *within* Australia?

I'm one of the producers who sticks to these rules, but they're constantly breached. I receive letters from customers saying they've forwarded our garlic to relatives or friends in the forbidden states.

When Graeme begged those thirteen garlic bulbs from a stranger in Kyabram, he distributed the bounty among family and friends, hence my gift of just one bulb. I couldn't know then how it would change my life. I've never felt particularly lonely at the farm, and I've felt less lonely still when putting address labels on boxes of produce and sending them off to people I will never meet. People who write encouraging notes with their orders are our extended family.

By 2010, within four years of our first crop, garlic had become our main operation at Elmswood. It has meant we've become retailers as well as farmers, stepping away from purely primary production. Surprisingly it's been a deeply comforting thing to do. Retailing takes time. It's time that would have been spent previously focusing on primary production. During the drought that began in 2017, when I did in fact feel lonely, it was our customers who kept my spirits up. Their friendly notes, comments about what we sell, how they used the products and plenty of tips helped the whole team to feel that what we were doing was useful and appreciated.

CHAPTER 2

CASUALISATION

MAKING A FARM PROFITABLE RELIES ON HARD WORK AND weather; even irrigators depend on the skies somewhere to secure their water supply. Nor can we control or predict productivity, let alone the yo-yoing of prices. It's always a gamble. Once in a while the rain comes on time and in the right amount, and you've a bumper crop. But bumper crops often mean slumping prices.

Industrial manufacturers, on the other hand, can control the inputs to their assembly line. They know that all their components will arrive more or less on time and be of a standard size, quality and price. And with robots replacing workers, they'll have no fear of strikes.

Farm owners are often the principal labour force themselves, but for agricultural businesses to grow, hired labour is essential.

Many of us still rely on the free labour provided by our nearest and dearest. It's impossible to place an exact price on what family agricultural workers contribute, but in our district, women's unrecorded work is critical. Children are part of the volunteer workforce too. Aurora spent her childhood doing myriad volunteer duties, including chipping Bathurst burr, spreading biological controls on weeds, turning pumps on and off, checking water tank levels, opening and closing gates to allow stock movement, picking olives, and, just before she left for university, harvesting, polishing and packing garlic.

Volunteering is said to be diminishing in Australian society, but I don't see it around here. Volunteers mow the lawns at church yards and keep cemeteries neat; they maintain memorial halls, ovals, sports clubs and the rural fire service. When there's a shortage of willing helpers for an event, call-outs are posted on social media and there's a quick response.

Inside and outside, farm work is industrious, and too many people over seventy still work long hours outside. Some might see this as elder abuse, but in this district if you can still work, no matter how old you are, you're a hero. You're also probably a first-responder volunteer.

But banking on the elderly to carry out farm work in the future? That's not the answer.

Young Australians today are mobile. According to the Australian Bureau of Statistics, more than half the people aged from fifteen to twenty-four changed their address between 2011 and 2016. They're not moving to Gundy. Most are heading in one direction – to cities. This would be great if it meant they were skilling up and returning home, but that's not happening.

With modern Australia founded on migrants, you'd think we'd be better policy-makers when it comes to immigration. You'd think that, as a nation, we'd have appropriate visas in place to help rural businesses access the willing workers they need. Yet visas that encourage farm workers seem to be an all but insurmountable political problem. Based on assessments by the Australian Farm Institute since 2010, the agricultural sector is short of approximately a hundred thousand workers. Hence many farmers are proposing core changes to the laws, such as introducing special agricultural visas, inviting workers from more countries, allowing workers to work on more than one farm.

I don't access foreign workers directly because we're not a big enough business. Elmswood can't accommodate large teams of workers, nor are our crops sufficiently large. But I like the idea of freeing up regulations that restrict holiday-makers from doing some farm work, even if statistics suggest that a holiday-maker is a less productive worker than a regular employee. For those of us struggling to get workers, a holiday-maker would be good enough.

If we don't sort out the issue of visas for migrant workers, the future of Australia's horticulture industry will be in great danger. Health reports for the past two decades have been recommending fresh fruit and vegetables as the number one thing we should all be eating, so horticulture should be centre stage in our focus on healthcare. But politicians prefer photo ops when they add a wing to a hospital.

If we don't improve farm visas the consequences are clear: it's what's already happening in other countries, some of them the world's richest. In Italy, 120,000 itinerant migrant workers are farm labourers. Malians harvest fruit crops, often in mafia strongholds. Think of that next time you enjoy an Italian kiwifruit. It was probably harvested by a man from Mali, living in a cardboard hut with no electricity, a makeshift toilet and no access to potable water. He was probably an illegal, with no worker rights. Shootings aren't uncommon and murders have been reported. Being a farm labourer in rich Italy is dangerous work.

Some temporary European worker camps hold two-and-a-half-thousand people. Some municipalities provide emergency bedding, but leftover workers can often be seen sleeping along the roadsides near the farms. And these aren't undeveloped nations – this is business in a European country driven by a market that demands cheap food for global trade.

Australian migrant workers may not be doing it as tough as Malians in Italy, but a similar situation could arise here if we're not careful. In Melbourne a centre has been set up to help migrant workers – many of whom work on small farms – understand the laws, their rights and obligations.

Meanwhile, in October 2018, Prime Minister Scott Morrison suggested that farmers advise the National Harvest Labour Information Service (part of the federal government's Department of Jobs and Small Business) about their labour needs so that unemployed local people can fill positions. As if we don't already try to hire local people. Farmer rage vibrated through the vegetable rows of the nation.

The prime minister quipped that everyone on the dole who didn't take up a farm job better have a good excuse, but threatening to cut benefits misses the point: harvesting and planting are intense times requiring short-term employment. And these jobs don't always fit in with people's employment aspirations. Dedicated agricultural staff are needed for our billion-dollar industry.

Once, there were a lot of people with a wide range of skills working on farms, able to pull apart the most complex machine, repair it and put it back together, trap wild pigs, do a spot of welding, fix a broken pump. Now kids are leaving in droves and family farms are being replaced by corporate farming. Once, the oldies would mentor the young; now such intergenerational involvement is endangered. Trained, ticketed contractors are doing so much more work on farms.

Over a hundred years ago, equipped with just axe, saw, crowbar and packhorse, Elmswood workers erected fences across vast paddocks and up and down almost vertical hills. The ghosts of their fences survive – testament to skill and determination – weathered, withered posts and rails sawn from the ironbark trees that we find fallen and forgotten in the regrowth. Thirty years ago we had to start again, stitching the property together with barbed wire and electric fences. Like painting the Sydney Harbour Bridge, it's a never-ending task.

When we first arrived here the casuals I hired had previously been full-time farmhands, shearers, abattoir workers or mechanics. They came with years of working experience, mostly fifty-something people, tough and intelligent. We made contact via word of mouth and soon learnt that many had worked long ago on our farm, picking up hay, shearing, chipping weeds, moving stock. Now my casuals are younger, often without any experience at all – not just of working on farms, but working anywhere. Official numbers put the casual workforce across the Australian agricultural sector as being made up of 25 per cent overseas backpackers – young people having a working holiday.

Farm labour is considered low-wage, low-quality work, a job for the jobless, but so many tasks require observation, thinking and physical fitness. Weeding garlic may seem like a menial task but to do it well requires skill, and the ability to work in a team. You need to know how to look after your hands, your body. For a few years I was able to hire two thirty-something locals to help weed and harvest garlic. One was a tiler by trade, one a former abattoir worker. Both had entered the causal workforce unexpectedly and unwillingly. They had no ambition to weed garlic and often didn't turn up. Their work was casual in both senses of the word. But there was no getting around the fact that casual labour was going to be essential to the future of our farm; there isn't a developed country in the world whose agriculture is not now dependent on them.

Agricultural job surveys in Australia are oddly clustered with fisheries and forestry (no wonder the old Shooters Party became the Shooters, Fishers and Farmers Party). A survey in March

2018 showed a 10 per cent increase in jobs from the previous year, to 324,500, across the three sectors: agriculture, fisheries and forestry. This was a small part of the roughly 12.4 million Australians in employment overall, and people were counted as employed even if they were working only on the day of the survey, and they had to be working for just one hour. Australia performs well against the OECD average and appears to have low unemployment rates. What we have in truth is a continuous decline in wage growth. It's flatlined, and is even lower than before the global financial crisis.

Workers had been at the forefront of my mind for some years when I learnt, in 2013, that a farm labour business had been set up right here in Gundy. I'd known the owner, Lindsay, as a primary school kid who'd go yabbying in our biggest dam and leave a few wriggling in a bucket by the back door. The barter economy. I hadn't seen him as a budding entrepreneur, but his new business was booming. He was tracking ripening fruit around the country and putting together ever larger gangs of workers.

Lindsay saved the day for me that year by mustering a team of local casuals via Facebook, sending an eclectic collection of workers to help with weeding and harvest preparations: a bartender from the local RSL, a woman needing extra money to pay for her wedding, a kid from France, and a 23-year-old parrot breeder from up the Pages River.

The following year, with Lindsay busy in South Australia, I called a wine grower in the Lower Hunter to locate experienced workers, but all were occupied pruning vines. Eventually, thanks to a vineyard reschedule, I secured a team managed by a contractor who'd harvested olives at Elmswood fifteen years ago. I met them one morning in the garlic patch, where we were harvesting. Two women and four men. Three were from Thailand, three from Myanmar. Four were Muslims, two Buddhists. Only the supervisor spoke English. Adding to the communication difficulty, all had their faces covered with assorted scarves, balaclavas and large-brimmed hats, with only eyes and noses visible. The anonymity was unnerving but I could see a hint of smiles beneath the camouflage, although mostly I sensed nervousness. I communicated by mime and they all worked hard. Arriving punctually every morning just after daylight, they'd walk to the bank of the Pages River and toss fishing lines into a large waterhole, hoping to catch a carp for dinner. They tied their lines to a she-oak, then checked them at lunchtime. If lucky they'd plop flapping fish into their eskies, rebait the hooks and put the lines back in.

They added to their dinner menu by loading buckets of green garlic scapes into the boots of their cars. No need to explain the virtues of garlic to these casuals. I'd watch them show the garlic to

each other with approving nods and sign language. The lunches they shared were an Asian banquet of the previous night's leftovers, served and enjoyed on towels and grain bags in the shade of a tree and washed down with cold green tea from large bottles. I never got to know their stories, but one thing was certain – they knew how to eat.

I began to wonder if Graeme was the ultimate casual worker. Like Phillip, Graeme had no formal education beyond the age of fourteen; his knowledge and skills came straight from life. He could work out the maths of a crop in his head. One night at dinner he admitted he'd never read a novel in his life.

Aurora, then fifteen, left the kitchen and returned with Zadie Smith's *White Teeth*. Graeme read Zadie for two weeks, more and more enthralled. 'Amazing,' he'd murmur. 'Amazing.' We felt like dropping Zadie a line to tell her she had a new fan, a 65-year-old farmer near a little town called Gundy. Oddly, it took longer to get him interested in films. 'Why bother?' he'd say. 'They're all made up.'

Graeme was living more often in what we now called the Green House, a cottage beside the Elmswood homestead where my mother, Thelma, had lived the last years of her life, with Aurora crawling, then toddling, around her lounge room. In those happy days we called it Thelmswood. It had been, and is again, used for visitors, backpackers and temporarily homeless friends. Despite his preference for sleeping in his matilda on the floor, Graeme had been persuaded to move into one of the four bedrooms. His swag fitted nicely on top of a bed.

I'd see Graeme sitting on the verandah of the Green House, smoking a roll-your-own, listening to country music on the radio while reading one of his beloved classic-tractor magazines, or the classified ads or the weekly prices of produce in the back of *The Land*. He'd look up as I passed and we'd give each other a wave. Eventually, due to requests from overseas backpackers, we put a TV in the Green House, connected it to Foxtel, and everything changed. With a remote control, Graeme became a channel surfer and discovered MTV. Soon he was an authority on pop music, fell for Lady Gaga and had a passion for Pink. But he remained loyal to his first love, Toni Childs. Such was his devotion to her that he'd often roll up his swag and drive 600 kilometres to a concert, taking her a box of garlic and a bottle of olive oil. 'Toni's into organics,' he'd say. Touched by his attention, she'd autograph CDs for him, which he'd play all the way home. And then play loudly in the garlic shed.

We always had music blaring as we packed the orders. Before the arrival of the iPod we brought CDs to play and argued about the playlist. 'No bloody Bublé,' the blokes would yell. 'No bloody Beatles,' Graeme would insist. 'No bloody Childs,' everyone would chorus.

Then Graeme bought a smart phone and practised texting during smokos. With his large hands it wasn't easy, but eventually he became a messaging enthusiast. Both birdwatchers, he and I started using our phones for identification and research.

If Graeme thought he'd heard something of avian significance he'd text me. *Did you hear that? Black cockatoos to the north?* Many of our texts were about feathers. The annual arrival of the oriental dollarbird, with its bright orange beak, was always a subject. Was it about to nest, or simply rest and refuel before heading further south? The eastern koel also triggered numerous texts but it was an unwelcome visitor, with its nocturnal squawking from the top of the *celtis* keeping us awake.

I eventually bought Graeme a bird book to add to his one-volume library, to help identify the hundreds of residents and visitors. After the digital world, a book seemed so old-fashioned.

Betty is another Elmswood worker I'd love to clone. With more and more casuals working here, I had less time to assess skills and personality. I couldn't be with every worker every minute of the day, but most of the time Betty was. She's a natural-born supervisor, because deep in her heart she likes humanity and is one of the least judgemental people I've ever met. The fact that she genuinely likes people helps make the Elmswood's teams of very different people workable. Betty is my 2IC, and whenever she talks about retiring I suddenly find I have a hearing problem. She's worked on our farm for more than ten years, spending long hours in the garlic patch and handling literally tonnes of these bothersome bulbs, from planting to packing.

She grew up in the Richmond area, outside Sydney, where vegetable growers became turf growers and where a horse could munch in an oversized backyard. Her father took her to local horse events and Bible classes. At seventeen she was a bank

teller, at twenty married, at twenty-one a mother. She and her husband Steve live in Stewarts Brook, a shadowed and pictur-esque valley 15 kilometres away. The brook begins its flow high in the Barrington Tops, and Betty's cottage is perfectly placed on river flats close to the old local hall. Stewarts Brook is a secret place, with perhaps fifty houses tucked into nooks along its solitary winding road. Long gone are the days when it was home to prospectors seeking gold, and the townsfolk are now a mix of full-time and part-time farmers, miners and retirees. For Betty and Steve, the Brook has been the perfect place to live with their beloved horses.

According to 2018 statistics, regional women have a lower rate of participation in the workforce than men, 56.5 per cent compared to 66.8 per cent. And it's the mature women who are falling further and faster behind. There's just less work around for older women. Yet there's always work for a fit, ribbon-winning horsewoman like Betty. What she has always wanted is flexible employment, a job that fits in with family life. Waitressing, nursery work, catering, garlic growing have all slotted into her working life, and the thing all her jobs have had in common is teamwork. Betty worked in the gig economy before it was called that, but unlike many today, she chose to be part-time and casual. More and more people are finding themselves involuntarily and unhappily in part-time employment.

While deaf to Betty when it comes to her leaving Elmswood, I'm now told to keep a sharp ear on AI. Seems I'll soon be hiring robots. It's odd hearing industry leaders or politicians say that the agriculture of the future will be based on innovation, as if the present agriculture system isn't. As if the business of farming hasn't been growing with the internet and with new ways of generating sales. Farmers are old-fashioned and modern at the same time, and are already using drones for mustering, robots to check fruit in orchards, and smart technology to manage irrigation systems.

That said, it's very hard to see how caring for animals could be handed over to automatons, and could a robot properly grade a garlic crop for colour and quality, for visual appeal? Will a robot ever be about to delicately remove a two-day old bee larvae and place it in a queen cell? Will a robot light my hive smoker? Will it identify automatically which of my different olive varieties need harvesting and take them to a processing plant? Or repair the leaks in a drip line caused by the piercing teeth of a fox? It's not over yet for human farm workers.

Just because a farmer sits at a computer desk doesn't mean the farm's more profitable or the work more interesting. And while innovation and technology are claimed to be the key drivers for improved farming practices, bigger tractors and better equipment are immensely expensive. Hardware, software – it's only the big operators who can handle soaring input costs. Smaller farms are locked into using second-hand gear and being very DIY.

Getting ready for a garlic harvest one year, I was distracted by the impending arrival of a few visitors – three hundred, actually, as part of a fundraiser for the National Trust of Australia.

The drawcard is Phillip's art collection: three thousand-plus pieces accumulated throughout his life and that now crowd both our home and a large barn. They're primarily antiquities, with an assortment of ethnographic sculptures, and loads of paintings, many gifts from the artists. The Trust promises it will be a short visit. A few cars and buses will bring the visitors; they'll meander, then have cups of tea and cakes home-baked by the local Timor Tennis Club as a fundraiser, with chairs borrowed from the Gundy hall.

Suddenly I'm thinking about tidiness, dust, spider webs.

Phillip had promised to do some rearranging and cleaning but didn't. In any case he doesn't notice dust. What did Quentin Crisp say on the subject? The dust doesn't get any thicker after the first three years. Phillip not only agrees but thinks the dust protects the antiquities.

My worry peaks three days before the event, and I recruit Betty to help in the house. In the kitchen I spot some spider webs near the ceiling, and with ladder-climbing helpers all busy elsewhere, I take off my boots and clamber from a chair to the tabletop. Reaching up with a broom, I teeter and topple. To avoid slamming my head on a kitchen island corner I turn in mid-air and hit the floor with all my weight on my left shoulder. The pain is like nothing I've ever known, and years later I still feel that fall, and fear it recurring.

When I scream it's Betty who reaches me first. She looks down at me, afraid and concerned, then calls an ambulance. It's a long time coming, as is the morphine. I'm incoherent, angry and agonised. At Scone Hospital they have to cut off my clothing. Soon a second ambulance transfers me to emergency at Maitland Hospital, where I'm left forever, but by now I'm so high I don't care about the National Trust or their three hundred visitors.

Nonetheless I make it home in time for the event, so drugged I feel quite hospitable. And I'm rarely calm on social occasions. With my arm in a sling, my main concern is that someone will bump me. Surprisingly I found that there's not only a doctor in the group, but three of them, and all express curiosity and sympathy and offer advice. One of them is critical of the delay in treating me at Maitland Hospital and predicts I'll need to get my arm rebroken and a metal pin put in. He's right.

We've not had an open day since. And the cobwebs can stay.

The episode makes me think about something that's become popular in rural areas, namely recreational visits that allow city folk to participate in farming activities, particularly harvesting, as a means of boosting farm income. I've had moments of enthusiasm for the idea but have never overcome the obstacle of providing all the extra services that visitors require. Toilets, first aid access, biosecurity notices, car parking, insurance, and my own presence as part of the package. Freeing up time to meet and greet visitors, in what is essentially a tourism sideline, is just too difficult.

Many little rural towns are now surviving by becoming theme parks, painfully picturesque with a chintzy cafe, a bookshop, an art gallery, a craft shop. But they rarely link to the farms around them. There may be wheat fields in the district, yet no flour miller in the town. Or orchards may prevail, but there's no fresh juice outlet. There's a detachment between town and farm that's as disturbing as that between farm and supermarket.

Ivy is another of our regular team at Elmswood. She came here straight after completing her HSC at Scone High School; it was her first job. Ivy doesn't have a driver's licence yet so relies on the school bus or friends to get to work. The combination of her nervousness and her time-poor mother, with younger children

to care for meant that driving lessons weren't feasible. Ivy can drive our bikes, buggies and trucks but only on the farm. She is far from alone in this. In our area many kids leave high school without their licence and, to my horror, some drive on the roads anyway.

Long commutes to workplaces are a frequent news story in big cities, but here in the Hunter, as in other rural areas, some kids have a ninety-minute commute to and from school. The drive to town or a major service centre can easily be two hours-plus and we think nothing of it. Although plenty of farm kids drive unlicensed on the farm, supervised on-the-road driving is hard to organise. I vividly remember the hundred hours of practice I did with a sixteen-year-old Aurora; rain, hail or shine, day and night.

Driving aside, Ivy is a natural horticulturalist, observing plant, soil and insects throughout the day. Bugs are captured for analysis, and photos are taken. Her young skills have been tested as a retailer at local farmers' markets, and as a farm photographer.

Judy, too, is a valued member of the farm's team. She has a remarkable ability to find things. When things go MIA – boxes, secateurs, screwdrivers, car keys, phones, spectacles, even Phillip's hearing aids – and all our major search parties have failed, Judy will arrive and, lo and behold, she beholds. She finds things lost in full view. She finds things left in entirely unlikely places. She also finds things under things, and down the cracks in couches. She finds, even more invaluably, termites.

Judy has remarkable balance. I'll discover her up a ladder I wouldn't dare tackle – changing a globe or adjusting the wall clock in the garlic shed for daylight saving. Circus Oz beckons. And she's a fixer as well as a finder. When a machine comes to a grinding halt, the cardboard guillotine needs a revamp, a whipper-snipper jams, the gas bottle malfunctions or small appliances expire, Judy fixes them. She's great at packing garlic and soap into boxes; checking weights and addresses and carrying out quality control. She measures our honey into 400-gram jars and oversees the inventory of jars, boxes, thank-you notes and labels. All in all she's a masterful multi-tasker.

In 2008 I was introduced to a new type of farm labourer, the wwoofer. This notion was born in the UK in the seventies, when an office worker bemoaned the fact that it was hard to experience farming life if you didn't know a farmer. Gathering a few friends together, she headed for a biodynamic farm, where in return for light duties they received free board and food and fresh air. When other people expressed interest, a system was set up: WWOOF, or Willing Workers on Organic Farms, also known as World Wide Opportunities on Organic Farms. People wanting to spend a few days or weeks on a farm would be introduced to farmers needing help.

Before signing up to host a wwoofer, I visited a fellow olive grower in Broke who'd been having wwoofers stay for years. I had a few questions. What quality of accommodation was required?

My friend's was comfy and generous. What food did she provide? The wwoofers ate with her and also had full access to her kitchen. What time did they start work in the morning? Eight a.m. Could you get unlikeable people to leave? She'd never had a bad wwoofer.

So I started taking wwoofers, and they opened up a new world for the farm. They came from everywhere, spoke a babel of languages, marvelled at kangaroos, played with our dogs, smoked the beehives, were scared by snakes. To date, around a hundred young wwoofers have spent time at Elmswood, as well as a few grey nomads and some high achievers taking time off from brilliant careers. Some ended up staying for months, absorbing the landscape, swimming in the river, falling in love with each other, becoming members of the family, and learning English while we practised a little German, Korean, Chinese or Italian. Laughing and singing (many travelled with guitars) as we ate together in the evenings. Working hard in return.

Our first wwoofers were brash kids who thankfully left after a couple of days. A few weeks later an email introduced two girls from Hong Kong. Could they come and stay for a week? Both were office workers and lived with their parents in a high-rise. They'd never left Hong Kong before. The farm would be culture-shocking but we welcomed them.

As they unpacked their bags it began to rain. And rain. There'd be no work that day, so we drove them around hoping to spot some wildlife. Stopping here to pat a damp horse, there to look at a soggy wallaby. And, at their request, we took them to church.

With no farm work to do, I suggested housework. But they'd never done that. That was their mothers' job. Could they cook? No experience, that too was mothers' work. What about washing the truck? They didn't drive. But their smiles made up for everything.

After that pair, the phone rang often from wwoofers and we were turning more away than we could take. Two German men came, both of them smart and educated. They'd met each other the day before at a youth hostel in Sydney. One was straight out of school, the other back from his second round of serving in Afghanistan. They were thinking about their future, making plans. There wasn't much German spoken during their stay as they spoke impeccable English. The elder of them amused himself by going out all night, hiking through the backcountry, navigating by the unfamiliar southern stars.

Soon after, an Austrian couple arrived and made traditional apple strudel, rolling out the pastry across the table, a metre wide and so thin it was almost translucent. They filled it with apples and pears before brushing with melted butter, and served it with a glass of fresh milk. It was like eating in *The Sound of Music*.

Wwoofing has to be one of the best ways to bring city and country people, nationals and foreign, together to share a little work, and for a few summers the phone rang every day with a wwoofer wanting to stay. We've had to turn wwoofers away these past two years as the drought put pressure on domestic water supplies.

Backpackers wanting paid work creates a totally different

mood on the farm. The two backpackers who came last were two French girls travelling the east coast of Australia in their wreck of a campervan. They arrived two days earlier than expected and were tasked with weeding. Like all paid workers, they preferred an eight-hour shift. The weather was cold and windy, the daylight hours short, and they were left to do the work on their own. If they'd been wwoofers they would have first come with me to check things, do a little in the garden, have a trip to town, then work alongside whoever was doing a task. We share the farm and all its ecology and culture with a wwoofer. Backpackers usually just need money and they rarely leave the house or paddock they're working in. Not once has a backpacker asked to walk the property to see the wildlife.

In 2016 a young South Korean man from the local abattoir came to wwoof during olive harvest. This was the first chance I'd had to get to know a contract worker who'd come to Australia on a special visa. It transpired that many Koreans had been hired at the abattoir; they were shy, nervous people who didn't speak much English and most sent their wages back to their families. Abattoirs are tough places to work, perhaps the toughest. After three weeks with us, our wwoofer headed to Newcastle to discover beaches, the abattoir a mere memory.

A couple of years later, in May 2018, a labour-hire manager for the abattoir was fined $43,000 for stealing from the pay packets of vulnerable Chinese-born workers, some of whom were full-time employees and Australian citizens. The Fair Work Ombudsman found that there had been deliberate exploitation of migrant workers, but the labour-hire company had conveniently gone broke in 2015. Little wonder that our South Korean wwoofer wanted a different rural experience.

Tighter food safety regulations have driven abattoirs to become specialised places. Our local Scone abattoir used to kill pigs, sheep and beef for local and domestic markets. Now they focus on beef, and bone most of what they kill. It's mainly boxes that leave the abattoir these days, not carcasses on hooks. Most meat goes overseas.

Being small businesses, farms depend on multiple contractors – mobile mechanics, irrigation specialists, agronomists, not to mention labourers. Atop the agenda of all Australian farming associations is the labour issue. It's one of the main reasons farmers join an association, to get help understanding all the rules, rights and regulations. Unfortunately, stories of exploitation are rife, particularly involving phoenix operators. A phoenix company is one that is liquidated to avoid paying wages, entitlements, taxes and creditors. Then, like the mythological bird, its principals acquire a new company, get a new Australian Business Number, and continue identical business activities under their new guise.

How could I tell if a company I hired through was legitimate or not? Phoenix activity is theoretically supervised by the Australian Securities and Investments Commission, but ASIC is not empowered to help workers who have been ripped off; its eyes are on the big players. Phoenix companies fly below the radar, moving around the country quickly, quietly, covertly and cleverly.

Anyone who's worked at the local abattoir will tell you about contract bosses arriving in groups, wearing khaki pants with lace-up boots, young women in tow. Gang-like. I've asked why people don't complain. They answer: Complain to whom? In industries run on dirty secrets, no one wants to listen.

And that's where we primary producers, inextricably linked to the slaughterhouse, are complicit. Knowing what goes on at the abattoir and not blowing the whistle is like knowing about domestic violence and doing nothing. When massive European immigration to Australia began after World War II, many towns were invigorated by the work ethic those men and women brought with them. Yet today's overseas workers are seen as 'work thieves', so few people care when their rights are denied and their entitlements stolen.

I was reminded of the importance of immigrant workers at the Tamworth funeral of a friend, who'd been born into the Italian community that grew tobacco and fresh produce. These families helped introduce locals to their vibrant culture and food and eventually were granted residency and citizenship. Their children, grandchildren and great-grandchildren are today invaluable members of the community. So how to reboot our rural worker shortage now? Let's start with an amnesty for all illegal farm workers. Clean the slate. It's an indictment of the times that the federal government has found it necessary to inquire into the need for national legislation to combat modern slavery, along the lines of the United Kingdom's *Modern Slavery Act 2015*.

Finally released in December 2017, the Australian report, called 'Hidden in Plain Sight', made forty-nine recommendations to combat slavery, estimating that 4300 people in Australia were indeed slaves, having been victims of human trafficking, debt bondage or forced labour. Chris Crewther MP, chair of the Foreign Affairs and Aid Sub-Committee that was responsible for the report, didn't mince words after his investigations: 'The appalling practice of modern slavery is a scourge that regrettably continues to affect millions of people around the world, including in Australia.'

So it's official. Australia may have moved on since the days of exploiting Indigenous workers on vast Northern Territory and Kimberley cattle properties, and the 'blackbirding' of Pacific Islanders for the sugar industry, but this issue is far from over. In June 2018, the New South Wales parliament passed the *Modern Slavery Act 2018*, and in December 2018 the federal parliament passed the *Modern Slavery Act* (Cwlth). Companies with a turnover exceeding $100 million now need to prove they're not part of a slave trade and to publish a Slavery Statement. We'll start to read these new corporate reports in 2020, so we're not really sure how effective they'll be.

We'll soon have an anti-slavery commissioner too. Let's hope these measures will be enough, and don't turn out to be like the pious but empty statements on sustainability. Because so much global food production depends on exploitation, we need these new laws to have a real impact.

For years our local farms have been robbed of young workers by the mines. Coal is the biggest crop in the Upper Hunter. During this latest coal boom, which began in the late 1990s, the mines in the Upper Hunter couldn't hire workers quick enough, or pay them enough. Who wouldn't go for the money? Farmland became vast open-cuts sucking the water from creeks and aquifers, mountains of overburden were created, and coal dust filled the air.

Farms lost not only their major source of labour, but also their skill base. Suddenly it was almost impossible to get good fencing done, or skilled mustering, or teams with the ability to make hay and silage. Now we're seeing the beginning of the end of coal, and hopefully with it our labour crisis will ease. But until then it's a problem for me, and it's also a problem for the coal workers. Once on full-time contracts with secured holiday pay and overtime, many are now hired on short contracts and have more in common with Uber drivers. This has changed the nature of the coalmining towns and communities too.

In 2015 I was in Des Moines, Iowa, on the day of a farmers' market, where local garlic crops were proudly displayed. Des Moines, too, had developed as a coalmining town, and every farmer I met that day worked their farm virtually alone or with family members. As in Australia, the 'volunteer labour' of family and friends was essential to survival. Few had ambitions to grow their business and hire paid staff. It was family farms versus corporate farms. From Gundy to Des Moines, the same global phenomenon prevails; farms are losing workers and, all too often, hope.

In many countries, governments subsidise agricultural income: in Norway the figure is 62 per cent, in Japan 43 per cent, in the EU 19 per cent. In the USA it's only 9 per cent while in Australia it's a meagre 1 per cent, though drought relief can temporarily up the ante. Around the world, various incentives heavily distort agriculture's declining share of GDP.

And industrialising countries are doing exactly what Australia did – using their natural-resource base to finance heavy industry. Farm workers are drawn to the magnets of mining or the cities, where there are better employment options.

John was another ex-abattoir worker who came to Elmswood, as a 33-year-old. At the abattoir John had worked in temperatures close to freezing, and after ten years he wanted out, literally. Out in the sun. We have plenty of that at Elmswood.

On his first day he stood with the new arrivals at the edge of the garlic patch. Most of them, John included, had cigarettes in their lips. Designating each a row, I gave a quick demo of the task at hand. 'Has anyone seen garlic grow before?' I asked. They hadn't. 'Do any of you eat garlic?' They didn't.

'Okay, this is garlic and this isn't. This is a weed, and so is this.' I named each weed and explained that the aim was to pull them out, not the garlic. 'And if you do pull out garlic by mistake, put it back immediately and press it into the ground.'

They smoked as they listened. 'Throw the weeds into the furrows, where the tractor tyres go up and down. The weeds will eventually die and turn into soil.'

John butted his cigarette and squatted down to give it a go.

The others followed. I tried to set the pace. Then, with considerable trepidation, I left them to it.

Like garlic, squatting was unfamiliar to my new weeders. Some people simply can't squat. Squatting has another meaning in rural Australia, of course, being the term for stealing land from Indigenous people. Those thieves later became Australia's rural aristocrats – the squattocracy – but there's nothing aristocratic about squatting in a garlic crop. That's much more down to earth, like squatting for a pee. Simply sitting on your bum and dragging yourself along is often the preferred technique. Either way, big hands help, and so do gloves. Kneeling as in prayer is also required, but many of the most devout weeders find that equally difficult. Even young knees find it painful, so knee pads, mini versions of those used in cricket, are handy.

John grew up in a big family, close to the Hunter River. The Hunter flowed behind his local school, so the river was his playground. He learnt to catch fish there, not only with a rod but with his bare hands, scooping out feral carp from the pools they'd muddied. He's so at home in nature and so skilled with his hands I think he'd have made a good early settler. He can spot a snake a mile off and is never remotely scared. His teacher would get the kids to look for snakes in the schoolyard, and showed them how to kill the creatures with a hoe – undoubtedly not a skill that was approved by the Education Department, but useful in the bush. And it taught the kids to be observant of their surroundings.

Although officialdom and the kids' parents might have disapproved, I admire that teacher. Yes, killing snakes is against the law, but sometimes you have no choice, Australia is home to most of the world's deadliest, and around here you have to be constantly on guard. Neighbours have barely survived being bitten by species regarded as only mildly venomous, and we've lost beloved dogs, a horse, and even bulls. We've found snakes in the laundry, hibernating under a TV set and swimming in the pool. John once scooped out a small black snake from our water

tank. Best to teach kids the protocols – how to identify them and how to react to them.

John's free-range childhood gave him skills that made him a better farm worker twenty years later. He learnt to be alert, observant, thoughtful, and those are the qualities I place at the top of my requirements for a farm worker. But it's not easy to find the people with them. Can we train them on the job – as farmers used to do with their children? I believe we can, and furthermore must try. How else are we to create fully credentialled employees? You'll need to like working outside. You won't want a regular day-in-day-out schedule. It's best that you're flexible and up for anything, whether it be plumbing or ploughing or loading cattle onto a truck. And on a farm, safety is all important. It can be as dangerous a place as an abattoir. Hi-vis shirts are useful (even I wear them), as are sturdy boots, tied-up hair, sunscreen, hats and gloves. There can be no speeding in vehicles. Beware of electric fences and badly behaved animals. Watch for snakes. Wash your hands.

The agricultural sector is coming to terms with health and safety, and with new technology – with the fact that expensive farm equipment needs smart people to operate it. And computer skills do matter.

When we bought Elmswood those thirty-two years ago I had to hire a farm manager. Reg, the man who'd been recommended to us, came for his interview with his wife, Yvonne, because if he got the job, both would live on the property with their three children. With any male farm manager, back then, you gained the entire family.

I remember that day as one of great tension. What on earth did I know about hiring anyone, let alone a farm manager? Reg had been highly praised, and that seemed good enough. And what should I ask Yvonne? I did say something asinine like, 'Is the manager's house okay?' and was assured she could help me with mine.

Many such people around here claimed rural heritage by dint of visiting their grandparents when young, or something similar. It's amusing hearing stories of connection, no matter how slim; there's a kind of competition on how rural your roots are. For me, a suburban Adelaide girl with no childhood friends in the country, there was simply no denying it. I was 100 per cent city back then. I had no idea about anything. It was Yvonne who changed that.

Yvonne was intimately linked to local history. She was born into a farming family not far from Elmswood, along Stewarts Brook, near where Betty lives and shadowed by Mount Woolooma, the looming hill I see from my office window each day. Her parents had stayed on the family farm until they passed away. Her brother still farms in the district, her children and grandchildren still live in the area working in rural industries. Together with their three young children, Yvonne and Reg lived on Elmswood for many years, until moving back to their own farm an hour up the road.

After Reg stopped working here, Yvonne came to Elmswood every week. She started out by helping manage the homestead, graduated to bookkeeping and we learnt together how to incorporate the GST, and gradually a computerised system. As the decades passed, she became my assistant manager and we discussed most business and employment decisions. Confidants as well as collaborators, we became each other's therapist.

Farming can be a lonely business. With Phillip away much of the time, having Yvonne as a soulmate made the farm, with all its dramas, more manageable. On the one hand a farm is columns of figures on a spreadsheet. On the other it's something living and breathing. Yvonne understood both. Especially the numerous small, precise, daily observations on a farm that measure failure and success, and they increased as Elmswood became more complex; in fact, no business demands more observation and collaboration. Cattle, sheep and kangaroos shared the property with olive groves, crops of lucerne and garlic, and beehives. Feeling a farm, sharing it with others, is what builds knowledge. I shared the landscape with Phillip and Aurora, with Reg and other managers, with many part-time workers. And I shared it all with Yvonne. We loved Yvonne. Everyone did.

In her world the man would work outside all day and return to a home where the washing had been done, the house cleaned, the food prepared. Her work was originally a type of servitude to a man's. I used to talk about this with her: as every working woman knows, women need wives too. But for Yvonne it was

the way of the world, and hers was a happy life. She'd ponder who was going to do all this work if not women.

Our weekly ritual that began when Reg left Elmswood lasted for years. Yvonne would come to the farm at 8 a.m. every Thursday. Around 11.30 we'd stop work for lunch. I'd make an omelette and salad and we'd eat out on the patio if it was sunny or by the kitchen fire if it was cold. It was a time to share confidences, problems personal and professional. We both had to adjust to the ever-increasing compliance issues needed to run a farm. She was managing all her farm's paperwork as well as mine, so we always had a point of reference. Learning how to do everything online with our slow internet was challenging for us. I have kept the old green leather ledgers from 1987 to 1999 in which we wrote out all the costs and income by hand. These look so simple now and forced me to improve my penmanship.

Yvonne helped me cope with the last nine months of my mother's life, after she'd been diagnosed with cancer and moved into the downstairs bedroom. I couldn't have got through that period without her. We had, however, very different personalities. Yvonne was patient and understanding, whereas I'm generally impatient and grumpy. I needed her patience and she enjoyed – sometimes – the anger I expressed. From childhood she had been taught to silence hers. When we met for the last time, it was Yvonne who was dying of cancer. She was sitting up in bed in the palliative care ward at Scone Hospital. A final farewell as friends and relatives gathered.

Knowing it was the last time we'd talk, we managed a smile about the secrets we'd shared with each other. All those stories told in confidence over nearly thirty years, things we thought we'd never share with anyone. At her funeral, bereft, I realised I was now alone with those secrets. I think one of the reasons rural people suffer more from depressive illness is that few have relationships like the one I had with Yvonne. Farmers must rely on their annual holidays, family birthday parties, and even funerals as therapy.

In 2017 the New South Wales government began the long process of designing a management plan for the Crawney Pass

National Park, which had been designated as a park in 2005. This area, part of the Liverpool Range and the Great Eastern Ranges corridor, starts just north of where Yvonne lived, at the head of the Isis River. It's the traditional country of the Wanaruah and Nungaroo peoples. There are nine threatened native animal species living there, and three species of plant needing conservation. From an ecological point of view, it is a crucial area.

Yvonne knew these hills well. She lived amongst them, loved them, cared for them. She'd be pleased that they're being given special status.

CHAPTER 3

WANTED DEAD OR ALIVE

TO CELEBRATE THEIR SUCCESS SOME LOCAL HUNTERS STRING dead dogs from trees: trophies meant to reassure locals that something dangerous has been destroyed – a crossbreed of a wild dog and a domestic rebel that had been doing its own policing of the hillsides. For a while the nights are quieter, the piercing howls hushed. But whenever I see a dog carcass dangling from a tree like a corpse on a gibbet or crucified on a fence I have to turn my head away in shame and revulsion. When I first saw one swinging in the breeze on my way to Scone I nearly crashed the truck. I headed straight to the stock and station agent, told him what I'd seen and asked what the hell was going on. He tried to soothe my anger by insisting it was necessary to alert the district to wild dogs on the prowl. It's not a tradition endorsed by government or wild dog management teams and with smart phones and Facebook the new bush telegraph it will, I hope, become a thing of the past.

There's a lot of death on farms, not all of it the killing of stock for food. At eight one spring morning in 2018, John sent me a photo of a dead sheep. One with our ear tag. One that had been shorn just three days before. Now it was a half-eaten carcass. 'I think a dog was on your lawn last night,' John texted.

I immediately forward the grisly photo to Richard Ali, chief biosecurity officer at the Hunter Local Land Services, commonly referred to as the LLS, in Scone. He calls me seconds later to say he'll come out. John is already on an all-terrain vehicle rounding up the rest of the flock. The wild dogs have scattered them along the dry riverbed and there's no sign of the lambs. Because of the drought, we've been leaving the sheep to wander what's left of my garden and the closest paddock, where there's still some grass near a water trough. Squire and CJ, our two dogs, were barking wildly at three this morning. I'd assumed their dog tantrum was triggered by the feral pigs we've been trying to shoo from the garlic crop. I'd been down in the paddock at nine the night before with a torch, shouting and stamping my feet like a lunatic and waving the light at two biggies with eight piglets running, snorting and squealing among the bulbs.

Richard spends the day with John reviewing the property and the problem. As head of the biosecurity team, Richard knows wild dog behaviour and is skilled in tracking and trapping. He knows how to find their dens, manage surveillance cameras, and, most of all, how to mimic their howl and call them in. He and John set up camouflaged cameras that snap photos silently and use a black flash. The sensor mode is triggered by movement across the lens space, and in between the movements of animals and birds the cameras shut down. Rechargeable batteries keep them going, and every few days the computer chips are checked. Richard has back-to-base access on some of his cameras, so he can check feral animal activity from anywhere.

I tell him of a yellow dog I've spotted some nights near the olive grove – probably the one that tore the sheep apart. Richard returns a few nights later with thermal binoculars, and a thermal-scope rifle with a range of 1.8 kilometres. He shoots the dog. But not before witnessing an extraordinary sight: two packs of dogs having a fight to the death, a territorial dispute on a hill close to our house. Seven dogs in all. The animals we'd been recording on cameras were just a few of a growing population of dozens living their dog lives across the farm, without us having any idea.

Wild dogs are like rabbits or pigs. You can try to keep them under control but you'll never get rid of them. Some are big and dangerous, not dingoes, but crossbreeds, and look like our pets. In areas like Mount Kosciuszko, the trapping of dogs has been an urgent issue for decades, whereas here in the Upper Hunter feral dogs, until recently, have been seen as an occasional annoyance. Now the state government funds the Professional Wild Dog Controller Program, to coordinate all local wild dog management plans. When Richard advertised three biosecurity jobs he received seventy responses. Each applicant had their own equipment and was already trained in trapping and surveillance. It seems there are plenty of animal vigilantes in the private sector ready to join the dog army.

Richard and a recruit return to check on the cameras, all of which have photos of dogs. Every dog is given a code name, usually a satirical connection to a member of staff at the Local Land Services office – office psychology LLS-style. Richard's personal working dog is a tri-coloured kelpie and a tracking specialist. A few days later I go out with them and the kelpie sniffs, scratches the dirt, points and tracks. We find dog scats and paw prints and Richard works out the direction the wild dogs are moving. They can cover 20 kilometres a night. Richard returns a few nights later and kills them.

Throughout our national parks and across some private land, when shooting is not enough, 1080 baits – sodium fluoro-acetate – are laid. This poison has been registered to use against feral animals since the 1960s, but as with most poisons there's no way to ensure that it finds only the intended target. The most humane way to kill an animal – if there is in fact a humane way – must surely be a bullet, because at least wild animals or eagles eating the carcass aren't likely to become collateral damage, as the military would say. But across the country, the LLS spreads tonnes of 1080 poison every autumn and spring, and we frequently hear about domestic dogs being poisoned. If you want to really annoy a farmer, kill their pet.

As someone who'd previously paid only casual attention to

wild dogs it was a shock to learn that so many of them were calling Elmswood home. When Richard discovered a pack of dogs living off our wallabies in a den up a gully, there was no question as to whose side I was on. Richard moved the cameras, we turned off a water trough so that the dogs would need to drink elsewhere, and watched our biosecurity battle between wallabies and dogs unfold until the dogs were gone. All dogs killed as part of this wild dog program are DNA tested. What often looks like a pure dingo can been found to be only 50 per cent. No pure dingoes have been DNA identified. It's true that when dogs are killed it changes the pack dynamics of animals in one area and opens up the territory for other species like foxes. We're not eliminating feral animals: we're trying to control them, so that the domesticated animals and the native wildlife on Elmswood, which has a much longer generational connection than us can be preserved. The death of indigenous wildlife feels different from sending an animal to the abattoir for the purpose of human food consumption.

When wildlife protection, environmental laws, animal welfare and farmer rights merge and overlap profound ethical issues remain. Of them the most morally challenging is theriocide, the act of a human killing an animal, any kind of animal. Innumerable acts of murder are deemed necessary or admissible, whether it be for sport, the putting down of an ailing pet, a setting of a rat trap or taking stock to an abattoir. The end result is the same. Death. We farm with a licence to kill, in accord with the beliefs and practice of our given culture. And those beliefs and practices are constantly changing. As have, for example, such contested human practices as abortion and capital punishment. The issue of theriocide is never far from our minds.

One day John found dead eagles in Middle Paddock, and when he told me about them my heart skipped a beat. He took me to see them. Two wedgies lying in the grass, about 3 metres apart. It was so strange. The most majestic of birds, eagles mate for life, and live high in the trees in nests as big as helipads. Was this a pair that some trespasser had shot? In the bad old days, farmers trapped and shot eagles to prevent them taking lambs, and then, as with the dog carcasses, hung them on barbed-wire fences as a grim warning to others. Or had these eagles been killed in some territorial dispute?

There was no evidence of foul play. They were just dead. Kneeling down, I realised I'd never been so close to one of these birds before. I could touch an eagle, examine the feathers, the powerful beak, the claws. For years I'd marvelled at them flying high, circling in the thermals, often in pairs, sometimes settling in a dead tree or on a fence post near a dead animal. Now I could pay my respects. Even in death they were not diminished. Two vast wild creatures, their eyes still open.

Little wonder that eagles are an ancient and universal symbol of imperial power – from the armies of Rome to the official seal of the President of the USA. A double-headed eagle appeared on the coat of arms of Russia's tsars, a use that was revived in the 1990s. The Greeks related the eagle to Zeus, and, according to legend, an eagle flew on Muhammad's banners.

When I had my first drone-flying lesson, trying to lift and

land the fragile thing on a picnic rug, the instructor asked about eagles. We scanned the skies and couldn't see any. But no sooner had the drone begun to hover than the eagles came, preparing to attack the mechanical intruder. That made me love them even more.

We didn't bury our dead eagles. Instead they became a pilgrimage site, until an animal got at them, scattering and mingling their last remains. I gathered the final feathers and took them home, later arranging them like flowers in an old jug on the patio. I've since learnt that some farmers still use strychnine to bait and kill eagles. Had I known at the time of these deaths, I'd have sent samples away for analysis. Now no evidence is left.

During the following months Richard came back to Elmswood to help with wild pigs, then rabbits. We had rabbit holes galore in sides of gullies, where I'd never noticed warrens before. Myxomatosis, the disease spread by the myxoma virus that brought the rabbit plague under control in the 1950s is still around, but Richard and his team now spread the latest calicivirus. We ordered some carrots to be used as bait and Richard chopped them up in a machine and spread it out in rabbit like scratchings, 30 cm long by 8cm wide. Day one, it's just carrots to tempt them and once they appreciate the free food, the calicivirus is mixed with the carrots. Day two and there isn't a single rabbit in the vicinity.

While I've only ever seen a few deer at the back of the property, we're now seeing increasing evidence of them. Antler and rubbing marks on trees tell Richard that fallow deer, and red deer, are now sharing the hills. These are considered game, not feral animals, a special classification to please professional and sporting hunters.

The New South Wales National Parks and Wildlife Service (NPWS) is the department charged with the immense responsibility of looking after state public lands. The staff are the keepers of secrets of millions of hectares, much of it in the wildest and least accessible landscapes. In those vast areas, the park rangers (parkies) are the only people paid to protect them from fires that break out because of foolishness, arson or lightning. Parkies also conduct planned hazard reduction burns as part of their conservation work. We've a state-wide Rural Fire Service (RFS) – I'm secretary of the Gundy branch. Far from amateurs, we're a 70,000-strong volunteer group working with 2100 brigades and around 700 paid staff. The division of duties may seem simple – the NPWS focuses on public land, the RFS on private land – but bushfires ignore bureaucratic regulations. Many fires must be fought by both services, the yellow team (RFS), and the green-and-yellow team (NPWS). When fires become serious, requiring strategic planning to control, the boss at a fire is always from the RFS. This is an unusual situation – the volunteer being allocated more power than the professional. The NPWS staff are not only paid but get a lot more training.

Our parks are public lands for all of us, mostly for pleasure, but occasionally as a haven for the notorious. Outlaws can be local celebrities too, not just the nineteenth-century equivalent.

Their hide-out caves are points of interest, their family burial grounds are noted in local cemeteries and highlighted on tourist brochures.

Malcolm Naden was suspected of murdering a woman in Dubbo in 2005 and evaded capture for seven years in the Barrington Tops, living off-grid. A real live outlaw living in our hills. Not even the $50,000 reward helped in his capture. Naden would rob local farms and soon had fans leaving food for him near the Barrington's dingo-proof fence. A local cafe served Naden burgers.

Police set up the same cameras we use to track wild dogs, and Naden was eventually arrested near Gloucester. He was tried, found guilty and jailed for life. His ability to survive in the bush for so long says something about the scale of our national parks.

All government departments suffer budget constraints and have to manage with fewer resources and people, but in the case of bushfires this problem is worsened by many NPWS elders taking voluntary redundancies. Decades of experience and knowledge of landscape are lost. And as elsewhere, many full-time positions have been made casual or contracted.

The police service is another one to suffer resource losses. Fewer and fewer stations are now staffed in the bush. Needing a document signed, I learnt that both the Scone and Murrurundi police stations were closed for the day. It's the same with NPWS offices: you can phone the local office and get transferred to a voice-message bank in Barrington.

While firefighting is the number one job of the NPWS,

they're also at the forefront of the biosecurity wars, aka pest management, doing the same work that Richard does on private land. A lot of hostility accrues to the NPWS. They cop the blame for everything from fires to weeds to feral animals. There's a local saying that if there's a wild dog around it's wearing a National Parks collar. But private landowners tend to forget that it was their predecessors who brought the weeds and the animals that turned feral into the country. Private people living on private land.

Soon after becoming a full-time employee at Elmswood, John completed his RFS training and was allocated a new outfit: high-quality boots, lightweight yellow trousers, a shirt with loads of pockets, a jacket, and a helmet with a visor and protection for the neck. I've only been at the frontline of two big fires, and at the first one I wore my own clothes, just sensible trousers and a long-sleeved shirt. Some things do change for the better. These days a rural firefighter isn't allowed on to a site unless they're fully trained and wearing appropriate official clothing.

The last fire close to home was at Christmas in 2017, when lightning struck a nearby hill. It was steep terrain with no designated tracks to the fire site. Phillip and I were in Sydney at the time and received phone calls and text messages asking about forgotten tracks and a creek I'd never heard of, which was coming up on the GPS. We headed straight back to the farm. Despite the fact that the official record wasn't matching local knowledge, the Gundy fire brigade and the NPWS had found their way to the scene. The fire simmered for weeks, with huge tree roots burning underground. Not a drop of rain fell to help finish it off.

At the annual general meeting of our local brigade, six months after the fire, we were still discussing who'd had control on the ground, how more training was necessary to liaise with the helicopters, and how the bulldozers had come too late to make firebreaks to stop the flames spreading. Every fire educates us but when lightening hits and ignites a hillside it's like God sending a message. No preparation is ever enough. In a second you can be in mortal danger. Dead or alive.

CHAPTER 4

NEW IDEAS

BY THE START OF THE TWENTY-FIRST CENTURY, HARDLY A DAY passed without more evidence of climate change, yet Australia did not become a signatory to the Kyoto Protocol until 2007. Earth, our living, breathing system, was and is in danger of being suffocated.

The Millennium Drought drove a local campaign to block a coalmine from being gouged at the head of the Pages River (a story I told in *The River*). Finally, in 2010, we won that fight, but there were other battles. How to manage the farm as climate change intensified; and how to engage more effectively, politically and practically, with this national and international problem.

In the climate change debate – as far as we've had one in Australia – the focus has been on energy, primarily electricity supplies. But there's one area that rarely gets a mention, namely bioenergy. In Europe, the US, Brazil and North Asia, on the other hand, it's a topic of immense importance.

Bioenergy is an energy system that engages deeply with the earth. It is genuinely renewable, and wholly remarkable. It provides fuel, gas, heat and power from plants, animals, their by-products and waste streams. It already provides around 10 per cent of the world's renewable energy, but in Australia it's estimated to contribute between 1 and 3 per cent.

There have been plenty of government reports over the years to confirm the benefits of bioenergy, but it's been relegated to the footnotes of the energy debate. This is partly because every bioenergy system is unique. If its economic advantages could be sold to every marginal electorate in the country, the sector would have more leverage and research funds.

My own enthusiasm for bioenergy was fired in 2006 by the possibility of keeping carbon in the soil – the fact that we farmers could manage our land in such a way as to remove carbon dioxide emissions from the atmosphere. At that time, Joe, a fellow sustainability advocate with a background in big business as a chemical engineer, was a crusader for an alternative energy to fossil fuels and had set up a company to develop a continuous biomass converter. This is a system for the slow pyrolysis (thermal decomposition) of biomass. Despite the modern name, this process is ancient – in the forests of Japan and elsewhere in Asia, and right across the Amazonian basin where forest timber was cut, plant mass was converted to charcoal and used to build soil.

When organic material breaks down on top of the ground, as in your garden mulch and leaf matter, the carbohydrates break down (photosynthesis in reverse) and eventually carbon dioxide is released back into the atmosphere. But if this matter, or biomass, is pyrolysed – subjected to extreme temperatures without oxygen – charcoal is created. (The word 'biomass' is used instead of 'fuel' in bioenergy, and is the technical term for any organic matter, which means all plants and all animals.) In slow pyrolysis the biomass is heated to around 400 degrees Celsius with virtually no oxygen. This process locks the carbon into the biomass, turning it into a form of charcoal. When buried in soil, this charcoal, or biochar, provides a stable form of carbon sequestration while helping to build soil structure and fertility.[10]

Slow pyrolysis in a continuous biomass converter also creates other useful products. First, there's a gas that can be used for energy, including to power the pyrolysis process itself. Second, a remarkable liquid is produced, an acidic substance called smoke water, wood vinegar, or, more accurately, pyroligneous acid. This also has an ancient history, as a fertiliser and natural pesticide. Thus, slow pyrolysis is so much more than just making charcoal, but it was the charcoal, the biochar, that excited me the most.

In 2010, after the third international biochar conference, in Rio de Janeiro, I took a boat trip up the Amazon, setting out from Manaus. There I found the future in the past, visiting famous charcoal sites and meeting farmers who were reintroducing charcoal into their systems. Archaeological studies have shown that life along the Amazon had been abundant, support-ing much larger populations than previously known.

The Amazon has triggered awe and wonder for foreigners ever since the conquistadors sailed up the mighty river in the six-teenth century. The river is so wide that it doesn't feel like you're even on a river much of the time, as you can't see land. Now, deep within the surrounding forest, more and more charcoal sites are being discovered. Some of the buried charcoal has been there for a thousand years. Scientists have dug pits wide enough to climb into, so you can see the layering of the charcoal, together with fragments of terracotta pots. This isn't a natural layering of charcoal after bushfires, it's charcoal made by indigenous peoples.

The fact that carbon can be made and stored for such immense stretches of time proves the worth of reviving this ancient idea to help remove atmospheric carbon. Energy production is slowly shifting to non-carbon sources (solar and wind farms are already cheaper to operate than coal-fired power stations),[11] but biochar produced by slow pyrolysis could help extract from the atmo-sphere the carbon dioxide of our past mistakes. This is potentially a huge game-changer.

That immense, dark, complex, impenetrable forest of the Amazon thwarted the European thieves and their discovery of the area's riches. It was hundreds of years before a new generation of explorers visited and saw the potential of the land as the next agri-cultural nirvana. After all, big trees in Europe were synonymous with fertile lands. But European-style farms and plantations failed to thrive because the Amazon soil, although capable of supporting huge trees and complex biodiversity, was functioning as part of a unique, whole living system. Tear this apart and its nutrients were inadequate. The making of charcoal by the indigenous people of the Amazon had been done on a small scale, integrated into their lives of growing and cooking food.[12]

Today anyone can make biochar with a small or medium-sized stove, or in a smouldering heap in the backyard, covered in dirt. On that scale it's not much harder than making compost. But the old way wastes the gas made in the process. The twenty-first-century version of the process can be scaled up or down, from mighty industrial operations to modest models for farms and communities.

In 2008, with private funding and government grants, Joe and an engineering team set up a research and development facility in Newcastle, next to an inlet of the Hunter River, to work on their slow pyrolysis, continuous biomass converter technology. As the technology was being fine-tuned, I gave up some work at the farm to focus on the sustainable application of this developing technology and, with the help of a team at the University of Newcastle, became one of three postgraduate students. Kerrie was working on the impact of biochar on soil structure, while Annelie was investigating the engineering requirements to improve micro-algae production. Algae can be grown in tubes or outside and there's plenty of research around the world now to develop it as a biomass source for many different industries.

Between the competing symbols of mountains of coal and a giant windmill, the engineers and students worked with passionate intensity. Some days things went well, on others there were unforeseen complications. That's the nature of R&D. Arriving early one morning I found a computer humming in the empty office and worried that there'd been a break-in. There were always concerns about industrial espionage, the stealing of secrets.

'Your computer was running when I came in this morning,' I said to Steve one of the lead engineers as he draped his jacket over the back of his chair.

'Yes,' he said, 'it's been working overnight on some maths for medical research.'

'Your computer is used for medical research?'

'I donate the electricity use.'

That's what I liked about the place. There was a lot of goodwill.

Most conversations I'd overhear during the day involved incomprehensible, technical jargon that needed an Enigma machine for interpretation. Or they might go something like this:

> 'Let's use a metre of pipe for here and that'll give us more flexibility.'
> 'There's not a lot of pressure here.'
> 'We can get a reasonable seal on this O ring.'
> 'This filter assembly will just clamp on top.'
> 'We need to go down there and work interactively.'
> 'Where should I put it?'

'On top of the hopper.'

'Put this on the outside of the flanges.'

'I'm not happy just having a hole coming out of the pipe.'

Some days it felt like working in the bicycle shed with the Wright Brothers. Other days it was like playing with Lego with kids.

The continuous biomass converter was a thermo-chemical technology feeding biomass continuously in at one end and combining functions of dewatering, char making, tar cracking (the breaking down of molecules to decompose tars) and gas scrubbing (the cleaning of gas), all within the one reactor. Coming out the other end in three separate outlets were gas, char and liquid. Joe and his team were seeking to determine whether this technology could rectify the techno-profitability gap that kills off so many bright ideas. And whether it would prove to be 96 per cent efficient in converting biomass. The unit being developed was designed to handle 10,000 tonnes of biomass per annum. The outputs would vary according to the inputs, but as a rule of thumb, for every tonne of biomass pyrolysed, 8 gigajoules of gas, 350 kilograms of biochar, and 400 litres of pyroligneous acid would be produced.

That system has not yet been commercialised. New energy systems don't get an automatic marketplace welcome, and there's a lot to consider with slow pyrolysis. Where was the biomass going to come from? Could the supply be guaranteed? What would it cost? Could it be supplied only from marginal land?

Should the biomass ever be irrigated? Were perennial crops best? Trees? Would there be enough local green waste? Could domestic garden waste be reimagined? How far should it be transported?

New technology like this is always risky and faces a lot of opposition from naysayers, because it's complicated and not understood. For slow pyrolysis to be accepted there needs to be a wide understanding of the full potential of its outputs: charcoal, pyroligneous acid and gas – three things that most of us don't use directly. And to start with, a sustainable supply of biomass is required, without which nothing can proceed. The gas produced in pyrolysis is made up of carbon dioxide and hydrogen, and is often referred to as synthetic gas, synthesis gas or syngas. It can be hooked up to a generator and converted into electricity or channelled into heating a building. The options are many, varied and complex.

According to Stephen Joseph, an engineer and biochar advocate, by 2018, China was operating at least fifty large slow pyrolysis plants and it has been estimated that they are producing more than 200,000 tonnes of char a year, along with gas for energy. The manufacturing of biochar-making equipment is also growing, with China leading the way. In Australia, industrial-scale operations are now coming on stream with the first commercial plant opening in 2018 at Tantanoola in South Australia. This plant, called Echo$_2$ uses biomass to make syngas and biochar.[13] Small-batch operations are proliferating. Some businesses sell charcoal to consumers who want a little char for their soil needs, or to feed to animals; many are gardeners.

Collectively this suggests that slow pyrolysis is off and running and that biochar will become more available and affordable. But because it's rural-based, and there are so many different engineering systems, it's hard to gauge the momentum.

One of the criticisms of bioenergy in general, and slow pyrolysis in particular, is that biomass production will ultimately reduce food production. This idea is a furphy. All natural landscapes provide multiple services. A suburban park is valued as a place to walk the dog, meet friends, have a picnic, but it's also a carbon sink. Chances are it's also pruned or lopped from time to time and is therefore a source of biomass. At present, such prunings make up the bulk of green waste, but they're

not really waste and shouldn't be seen as part of waste management. Anything organic (carbon-based) can be used as an energy source.

I saw how biomass could be integrated into existing agricultural systems when I travelled to Sweden in 2014 to visit Annelie, now graduated, who had moved on from laboratory experiments in Newcastle and was then working for an innovative algae company in Simrishamn, in a south-eastern region known as Österlen. As head of production she was farming algae in tubes, a whole greenhouse full of them. The valuable omega-3 oils in the algae are extracted using a solvent-free process and sold as a nutrient supplement.

The landscape surrounding the algae farm was intensively agricultural, with wheat crops and orchards sharing farmland with a stand of willow or *Miscanthus*, grown for biomass. These fast-growing, carbon-sinking crops would be lopped off and sent to a woodchip facility. Biomass is a valued crop on many Swedish farms; the trees also act as windbreaks and are often grown on less productive land, some in difficult terrain. The biomass at Simrishamn was not in huge monocultural paddocks, but part of a thriving patchwork of agricultural activity, integrated into the land management, so that at a glance you might not notice.

Because the source of biomass can be any organic matter, there are plenty of options besides council green waste. Household prunings, animal manure from intensive feedlots, recycled timber, even residues like pecan shells, which can make up 50 per cent of a nut crop and often end up in cattle and dog feed, are all potential biomass sources.

Crop residues, besides being a potential biomass source, already play an important role in soil protection, and on a small farm or plot they usually stay on site. My small olive prunings are potential biochar feedstock, but we mulch them back into the earth to become the soil of the future, while the large prunings are either used in gullies to help control erosion, by slowing the flow of water, or placed in the vegetable garden to raise the beds. Our garlic stems, both fresh and dried, are fed to sheep, and leftover bulbs are given as special treats to cattle. Cattle and sheep manure is converted back into soil with the aid of fungi and introduced dung beetles. Every time we sell cattle, we don't just sell meat, as an animal is also fertiliser (blood and bone), fat, leather, and countless other, smaller but important products. Residues on a farm are not waste, but biomass.

Regardless of its source, biomass needs to make the best use of that plant or animal material. It needs to be sustainable in both supply and price, and the use and price of the three raw outputs also needs to be defined. Slow pyrolysis could create plenty of downstream businesses, such as pelletising the biochar, selling the liquid as a fertiliser or green chemical. But in the short term, the three outputs need a primary market.

My thesis focused on the Muswellbrook local government area, the coal hub in our district. Today it is a wealthy rural shire with a strong economy based on coal, but that will eventually change as our reliance on coal changes. Already massive rehabilitation of the land where the coal has been removed is under way. These disturbed lands will never be the same as they were pre-coal, but they will still have some future economic value. Biomass production may be one. And biochar could certainly help redevelop many of the disturbed soils.

Biochar can satisfy any number of commercial needs to increase soil health and production capacity. As it can be manufactured from different feedstocks it would likely be priced differently, depending on the source. It's unlikely that there'd be a consistent wholesale or retail price until a consistent volume can be produced. The biochar price will also be affected by the price of all the other pyrolysis outputs, especially the gas product, which may or may not be able to earn carbon or renewable energy credits (a political factor), and the water product (the market value of which in Australia remains uncertain). So the price of outputs plus the primary energy saved will define the profitable price point. Presently prices vary from 70 cents a kilogram to $6 a kilogram for different grades of biochar. But with agriculture globally facing low profit margins, biochar will need to find a cheaper price entry point. And this will depend on related enterprises – horticultural, broadacre or industrial farming, land rehabilitation.

The International Energy Agency (IEA) has calculated that for bioenergy to produce 20 per cent of the world's primary energy needs by 2050, special biomass production will need to increase fourfold. Containing the transportation costs of biomass lignocellulosic (plant matter) feedstocks will be a challenge.

As biomass sources weigh less than coal, more transportation movements will be required.

The IEA has based these biomass requirements on technology conversion efficiencies of less than 60 per cent, whereas the continuous biomass converter has a conversion efficiency of around 96 per cent. Therefore, the true value and efficiency of biomass requirements could be better than the baseline projections used by the IEA.

Because investigations into the use of pyroligneous acid had showed that it could make biochar even more effective as a soil improver, it seemed important to do an experiment on the farm. So in 2013 with cowpeas, a green manure crop that we grow in summer and plough back into the ground to help build soil structure, we tested the impact of growth on biochar and pyroligneous acid.

The humble cowpea is a remarkable plant that fixes nitrogen as it sucks up the sun's energy. Although we call it a cowpea, as if it's for cows, it produces a delicious green bean that's a food staple across Africa, while the young leaves can be used in stir-fries. Our cowpea crops failed these past two years due to the ongoing drought, but in a typical year we'll plant cowpeas straight after the garlic harvest, to keep the soil cool and refresh the land. Nothing makes me happier than to see this crop waving bright and green under the summer sky while supplying endless beans for munching.

For the experiment, I used a river flat near the garlic crop, and tested biochar at 5 tonnes a hectare and 10 tonnes a hectare, and then added pyroligneous acid at 5 tonnes a hectare and 10 tonnes a hectare. I also had a plot with just pyroligneous acid on its own, without biochar, and of course there was a control plot where nothing was added.

The results showed that the addition of 10 tonnes a hectare of biochar had a statistically significant positive effect on plant growth, and that using higher volumes of pyroligneous acid could be applied without any detriment to plant growth.[14]

Weather always dictates growth, and the experiment was plagued with extreme heat of 44.5 degrees Celsius and higher, and low rainfall. I get annoyed when I hear people argue that irrigation is the be-all and end-all of crop success. A crop can receive the perfect amount of moisture during its growth phase and then be spoilt by a storm right at harvest time.

Adding more biochar to our soils at Elmswood is a future plan which will likely be in a small operation using our olive wood after pruning. Why waste waste?

As the char used for my experiment was made from sawdust, it was powdery, hard to handle and easy to get all over you. A mask, gloves and old clothes were essential. After being spread on top of the soil, it then needed to be mixed in. I would have preferred pelletised char. Cattle producers have found that adding char to their stock's diet, usually mixed with molasses, makes it palatable and their manure is a natural, easier way to spread char across a paddock.

E.F. Schumacher, in his important 1973 book *Small is Beautiful*, understood the difficulty each farmer faces in defining the proper scale of an enterprise.

But small isn't something I see when passing huge coalmines every time I leave the farm for Newcastle or Sydney. With the wider world acknowledging the need to phase out the use of fossil fuels, the pace of coal business is quickening in the Hunter Valley. The thinking seems to be: dig it up and ship it out before our export contracts fade.

At the R&D facility at Newcastle, we watched infinite amounts of coal arriving at the harbour in endless trains from the Upper Hunter Valley mines, piled ever higher before being loaded onto huge ships. Australia is one of the world's greatest exporter of climate change. I never fully appreciated the dominance of the mining sector. The city's business elites are from the mining companies, coal's multiple support industries and from the managers of the port.

The Hunter Valley may be Coal Valley today, but it could be the Carbon Valley of the future. Bioenergy could help meet our renewable energy targets, create economic development and jobs, provide greater diversity for the energy market, assist in land rehabilitation, and at the same time sequester carbon.

The progress of industrial agriculture has been based on new plant varieties, irrigation, synthetic fertilisers, pesticides, herbicides and genetic engineering, while ignoring the ecological degradation these things have produced. We've been able to do such massive damage so quickly because cheap energy has been behind the model. Bioenergy can be part of the alternative – a modern version of an old approach.

One of my favourite books, *Farmers of Forty Centuries: Organic Farming in China, Korea, and Japan* by F.H. King, first published in 1911 and since reprinted, provides extensive and inspiring examples, with plenty of photos, of how communities in those three countries have managed and protected their land and water for long-term use. It was in this book that I first saw a photo of commercial pear orchards where each pear was protected by a paper bag. I now do it every year with my pomegranates, otherwise the birds get them all. Farmers of a hundred years ago didn't talk about greenhouse-gas abatement, but that's what they were doing when they recycled the ash from their kitchens, collected herbs from the common land around their villages to make compost, and protected the soil in their orchards with cover crops or straw. Agricultural research stations were everywhere in Japan, Korea and China a hundred years ago, testing new plant varieties and equipment. Being innovative is old news. There are old and new lessons everywhere to help us mind the farm.

I don't believe we have a problem feeding the world's people, but we do have a problem with wasting food. Proudly displayed on a kitchen wall at Elmswood is a piece of propaganda, an old poster reading STOP THE CAPITALIST DIET. This isn't just an admonishment to avoid junk food, it's a warning about wastefulness. All food scraps go to the dogs, the chooks, our worm farm or the compost heap. We used to carefully separate tin, glass and plastic so that our contribution to landfill was small, and we'd truck it ourselves to the municipal tip, for want of council collection. Now with so much plastic packaging it all ends up in a skip we pay to have emptied each month (it's meant to be sorted offsite). Despite the best efforts of our dogs, chooks, worms and compost, and the War on Waste – which seems to be as successful as the War on Drugs and the War on Terror – it still feels like plastic is winning. Plastic irrigation drip lines have improved our water efficiency but given us a disposal problem. New pipe fittings are mainly plastic and arrive cling-wrapped with sticky plastic tape. Our old aluminium pipes are light, easy to store and reuse, but less efficient with water delivery.

Checking our dumpster I find bits and pieces of fencing, netting, weed mats, cans, the detritus of redundant irrigation tubing, and the ubiquitous blue twine we use for tying everything from hoses to hay bales. A friend infuriates the local supermarket by unwrapping any plastic-wrapped items after checkout and handing back all the plastic there and then. Could we all join her in this subversive act?

I try to match the amount of vegies I grow in the kitchen garden with what we can eat, so my rows of coriander are now shorter, the clumps of silverbeet fewer. Forget five zucchini plants this summer, two will be fine. We buy discounted fresh food, the ugly, the misshapen and the blemished. But even so, I must plead guilty to multiple charges of waste.

Figures from the UN's Food and Agriculture Organization

predict that the world's population will reach 9.1 billion by 2050, with 70 per cent living in cities. Most of the population increase will be in underdeveloped nations. Chemical and fertiliser companies want us to believe we're all going to starve unless we use their products in ever greater amounts.

Developers use statistics to argue that we must open new regions for agricultural development. Yet in Australia many fertile paddocks lie idle. The land is capable of storing a lot more nutrients and water, and therefore producing more. And this is possible without adding billions of inputs. The problem is a lack of people wanting to do this work, and a lack of people knowing how to do it. At the core of it all is a poor understanding and appreciation of our natural environment and its particular landscapes.

Land has a varying capacity to grow food, fibre, timber and fuel. Some years there's good rain and a bountiful crop, some years the crops fail. The agriculture sector is expanding in the developing world, while Europe and the USA are paying more farmers not to farm.

Every year there is a food surplus somewhere, which is kept from market in order to keep prices stable. Food storage is a sector all to itself. Meat can remain in coolrooms for years; grains in silos are pumped with fungicides to keep them safe from insects and falling prices; milk is powdered, vegetables frozen or canned, wool and cotton stacked in sheds.

In November 2017 the federal government launched its National Food Waste Strategy, with the ambitious goal to halve food waste by 2030.[15] Australians waste food just about everywhere, in the paddock, during transport, in factories, in supermarkets, in our kitchens. The Fight Food Waste Cooperative Research Centre claims that we waste 40 per cent of what we produce: that means we're also wasting 40 per cent of the water and fertiliser used in food production. Not to mention the trillions of wasted kilojoules.

Then there's the food we discard from our kitchens each week. The apple that went soft, the bread that went mouldy, the leftovers that didn't get reheated. Food waste is like leaving a coal-generated light burning, or a big-screen TV on in an empty room. Estimates suggest Australian individual households are throwing away $4000 worth of uneaten food per year.

It is impossible to tally the full amount of food wasted. My cows can get bogged in dams, or slip down a hill and break a leg and die – there's lost protein, fat and kilojoules, all unaccounted for. Our sheep can be killed by wild dogs (more lost meat), cockatoos eat the garlic, and pigs dig up my oats. All these are small contributions to that incalculable waste. Weather can spoil a vegetable crop, cyclones entire banana plantations, and there are also losses post-harvest: trucks spilling grain, destruction by weevils, food-processing failures, and waste in supermarkets. The national cost of this waste is estimated to be $20 billion each year. It's hard to see how we can claim to be a sustainable food-producing country with a statistic like that.

We demand instant gratification in everything, and most of all in our food. Accustomed to having anything we want to eat whenever we want it, we're indifferent to the energy wasted at every point in the food chain, along with the time, the effort, transport, and, most of all water. Agriculture uses 70 per cent of Australia's fresh water. When we waste food, we are wasting this most precious national resource. I'm always shocked when I see people toss uneaten food into public or private rubbish bins. I love the way a city neighbour of Greek ancestry kisses her bread if she needs to bin it. Imagine doing that to all the food you throw out, so many kisses of goodbye.

Then there's the fact that too many people are eating far too much food, and the wrong kind of food. The obesity epidemic is part of food waste too, and the money spent on futile diets is more than some people spend on food.

We don't need to colonise more land, use more water, spray more chemicals or use more fertilisers in order to feed the future population. We do need to stop wasting food and eating more than we should. Remember, too, that every piece of food that is thrown in a bin is destined for landfill, where it will rot away and emit greenhouse gases. Wasting food is bad for climate change.

In the debate about future food sources, there are some who claim that bioenergy will 'bio-fuel poverty'. That is, drive the world further into poverty, not help to overcome it. This argument ignores the fact that since World War II, more and more land has been taken out of food production, and that we need energy as well as food.

Farmers in Europe manage their land knowing they'll receive annual government subsidies, which they call income support. They regard this as a good thing, an acknowledgement of the fact that farmers face risks that no other sector does, such as weather and specific market variations. The European Union's Common Agricultural Policy sets the rules and regulations for what is grown and how much, what land is protected and what is removed from production, ensuring that rural communities are supported. Despite this, a good growing year can still lead to annual food surpluses, and a similar situation applies in the USA. I call on all of us to remember this the next time we hear the claim that we won't be able to feed ourselves.

I know no one in the biomass sector who wants to use whole fresh food for energy. That would be lunacy. It's crop detritus that can be used, all the inedible part of food crops, the fibrous or woody portion, the lignocellulosic material. The agricultural land of the future will ideally be multifunctional, with food, fibre, fuel, and carbon sequestration coexisting, as I saw in Sweden. It will incorporate marginal land, idle land, and less productive areas. This same land could be enriched with char and pyrolysis liquids, putting it at the forefront of carbon sequestration.

Bioenergy should play a major role in the redevelopment of rural areas and jobs growth, with the trade-off between food, fuel, carbon and fibre being unique to each landscape. But for this to happen, it needs to be embraced in sectors other than agriculture and energy; it needs to garner social support. Land-use changes in Europe and the USA due to the uptake of bioenergy have been driven by the need for heating and transport fuel, whereas in Australia rehabilitation of agricultural and mining land is our major task.

Back in 2008, the result of tests evaluating the warming of soil over the previous twelve years showed a decreased microbial biomass. If more soils are going to be hotter and drier for longer periods, what we can grow and where is destined to change, and more land is likely to be classified as marginal. The only really secure places for intensive agriculture will be in limited irrigation areas with high-security water access.

The future, in other words, does not require a battle between fuel and food, but a total rethink about land use.

CHAPTER 5

HEALTH: LAND, MIND, BODY

EVERY YEAR IN OUR SMALL NEIGHBOURHOOD THERE'S AT LEAST one suicide or attempted suicide, perhaps someone we've heard of rather than actually know. Some deaths are unequivocally suicide – bullet or rope – but there are others for which the coroner delivers an open verdict. A head-on into a tree on the Gundy Road, or worse, into another vehicle. A farm truck inexplicably driven onto the railway tracks. A fatal accident attributed to 'asleep at the wheel'. The official figures for suicide are way below the reality.

One morning in 2018 I woke to the news that a local man aged forty – not someone I knew well, but a friend to many – had shot himself. A few hours later I was discussing our wild dog problem with a feral-animal hunter and he apologised for not being available the next day, because a friend, a bloke of forty, had killed himself. I began to say that I'd heard the grim details when he added that his friend had hanged himself. We were talking of different deaths, two local men who suicided at the same time.

Across Australia, sudden-death statistics are dominated by domestic violence, by men being violent to women.[16] But more and more men are being violent to themselves. While more women attempt suicide, more men are successful. The male suicide rate is 70 per cent higher than the female. And that's before you factor in open verdicts, those coronial efforts at compassion. The phenomenon of rising suicide rates is particularly marked in localities outside greater capital cities, and worse in remote Indigenous communities.

Suicide has been part of my own family life. To a certain extent it can almost seem normal, but it shouldn't. We have an epidemic of suicide and we should not allow it to become normalised.

Few of us die of shark attack, even fewer from terrorism. Yet with the help of the media, both sharks and terrorists cause national panic. And so many more of us die from suicide than in road accidents. We spend billions trying to reduce the road toll, via ever safer roads and cars, sterner policing, higher fines, the demerit points system, driver education. But on suicide prevention? Mental health in general? A pittance. Particularly here in the country.

It's true that while the number of suicides has increased nationally, the percentage within our fast-growing population has decreased. Nonetheless the official figures for 2016 show that 2866 people chose to stop living that year, some as young as ten. Some of that total would have made the decision because of the pain of illness, but most kill themselves because of the pain of life. Not everyone leaves a suicide note. The two local men did not.

My mind keeps asking why those men killed themselves. Why so many in our district over recent years? Both of those men ended their lives without their family or friends suspecting that they were deeply troubled. It feels as if it's the community that needs therapy, to learn how to communicate honestly, openly. Though neither man was part of my circle, I feel a sense of failure at their deaths. So should we all. We should talk to each other, really talk, about our lives and issues, from the personal to the political. Instead we censor ourselves.

Employment issues often seem to be at the heart of rural mental health. With local kids desperate to leave home, towns are becoming ever smaller. The talent pool is as dry as a dam in a drought. Regional centres, too, struggle to exist. At the time of writing, the Geographical Names Register of New South Wales lists 261 towns in the state. In the Upper Hunter shire there are a total of seventy-four 'designated localities'. Gone are most of the actual villages – what's left is a ghostly memory of a place, the odd store if you're lucky, or a one-teacher school.

Gundy is a case in point. In the 1950s it had one of every-thing, including a row of stores that made it so photogenic it was used as the location for the odd Australian film. Then one by one the businesses closed and the stores were dismantled, like film sets. Now only the Gundy store and the Linga Longa Inn survive as private businesses, unless we call the Church of England a business. Our memorial hall is the centrepiece for

Anzac Day, an annual ball, weekly childcare mornings for local children; and the Gundy oval is used for public events, private campers and family get-togethers as well as for sport.

Many rural areas have community halls standing alone in beautiful settings, some recently renovated thanks to government grants supporting 'heritage', even though there's not a lot of community to keep them alive. Still, one of the most noticeable differences with drought relief in 2018 compared to previous droughts when rebates were more common has been money. The money allocated for community events, to cover the basic costs of food, music and hall rental: firm recognition that community fosters mental health.

In his book *Lost Connections*, which considers the treatments that are handed out for depression, Johann Hari reviews the growing evidence that some people are driven into depression by 'sleep working' in jobs they don't enjoy. I thought of the Hunter's coal industry when I read this. Many people in the area have sought jobs there, and once employed have become cogs in a mighty machine. They sit at the controls of a juggernaut in a coal pit, having no contact with other workers or with the land. It's easy to see how that's not good for mental health. Farmers aren't usually in this category of 'sleep working' – the work is too diverse, too physical – but tough times can provide tipping points. Especially drought.

Droughts are as unpredictable as the rest of the weather. Sometimes they arrive suddenly, sometimes by stealth, but there's always a point when it suddenly hits you. And when it does, I turn as brittle as the landscape. I've been dealing with droughts for decades, with the inevitable ten-year cycles, and in every drought the behaviour of our animals changes. They cluster together under different trees, create new tracks to the top of a hill. A cow will stop and look at me with its deep sad eyes as I walk past, both of us silently acknowledging to each other the harsh weather. The female satin bowerbird will stay perched on a cabbage when I pick herbs a few feet from her. A rock wallaby, usually the shyest of animals, will sit stock still as I pass just metres away.

The deepening thirst of climate change is unprecedented, as evidenced by this latest drought. One morning, as I walk naked (I'm all alone) to sprinkle some precious water on the vegie garden, I'm startled by a huge black shadow. A bull has jumped a fence somewhere and now just stands there looking at me. Not threateningly, just curiously. Later, driving out to check on things I see two eagles soaring above me. Another dead animal nearby.

At the bank, the supermarket or anywhere people chat on the footpath, I'll hear the gossip about who's completely destocked, whose pump has failed, who still has a dam with some water in it, who's on antidepressants. Overseas friends send emails about droughts in their countries, as if this will make me feel less alone. It doesn't. Much of Iraq has reoccurring droughts like Australia and there's now a Save the Tigris Campaign. ISIS is discussed – not our feeble little Isis River, now a string of brackish puddles – but Iraq's terrorist group. In Baghdad the Water Resources Minister announces that the water flow is down 40 per cent, due not only to less rain, but to the curse of dams.

Rivers may physically connect people, but with the first hint of drought they psychologically divide.

With so many suicides and the drought, there are times I yearn for the ease of city living. A trip to the Art Gallery of New South Wales in Sydney is perfect R&R. I check out the Wynne and the Archibald entries, whiz through the gallery shop for a few cards, then hike around the permanent collection looking at old favourites, like Drysdale's gaunt paintings of drought. Then it's back home again, not looking at images of drought in air-conditioned luxury, but feeling it, fearing it, seeing the ugliness. Especially on other people's properties, where too much clearing and gross overgrazing has deepened the impact. Up here you can be a respected local farmer yet be utterly brutal to your land.

Every five years, the federal government gathers data on the nation's environment and then releases a State of the Environment report. I remember being given a hard copy of the 2011 effort as bedtime reading. It was so heavy I needed a pillow on my lap. The colour pictures, pie charts and graphs told a shocking story: if ever there was a book to trigger outrage or despair, this was it. Yet what has changed since then?

We've also learnt a lot about institutional brutality and domestic violence, about child abuse, about the stolen generations, about Aboriginal deaths in custody, all of which have been the subject of inquiries, commissions and subsequent reports. So many government reports and so little follow-up action that addresses underlying causes. The New South Wales Department of Primary Industry has set up a Combined Drought Indicator portal to help, not just farmers but everyone, understand the history and scope of drought. Using detailed maps and charts, land across the state is divided into categories from non-drought, recovering, drought affected that is weakening or intensifying to intense drought.

This isn't to say that help isn't offered during crises. There's the state Department of Primary Industry Drought Hub website, useful for those with good internet connection. The Country Women's Association of New South Wales set up a Disaster Relief Fund for farming-dependent families and contractors who can prove they're struggling. Not-for-profits distribute hay and fodder. Drought coordinators pop up in rural towns. This is on top of the $1 billion of public funds given for drought relief. The team at Hunter Slow Food provide boxes of fresh vegetables grown along the riverbanks two hours' drive away, where rain still falls. Another group, Aussie Helpers, coordinates psychologists to be on the end of a phone to help people talk through their anxiety; they reported getting sixty calls a week in August 2018. But while there are many helping hands, that doesn't mean it's easy to reach out and touch. And it's too early to tell if the money and advice, medical or financial, will be of much long-term help.

Across drought-affected areas there are 'support the farmer' meetings at halls and pubs. They've been going on for months at Gundy's Linga Longa Inn. I find a flyer in the letterbox and doubt I'll want to linger long, but go anyway.

At the pub I hear about the latest forms of assistance for struggling farmers, who may own a multimillion-dollar asset but have no money in the bank; their capital investment has become utterly unproductive. Hand-feeding stock is wildly expensive. Most farmers at the pub that Friday night have other sources of income besides farming. There's an electrician, a builder, a schoolteacher, a dozer driver, a councillor, a jeweller, a student, a few government workers, a coalminer, a truck driver, rural agents, some staff from local businesses, an investor, a few pensioners.

I'm worried I'll say things I'll regret. The people at the pub are, like me, very stressed, full of inchoate anger; the veneer of politeness is thin. But there's interesting news too – about Rio Tinto's new Chinese shareholder, some tree planting at BHP's Mount Arthur coalmine, the whopper outside Muswellbrook. Who's just been hired at a mine and who's left. Not all coal mines are unionised, and full-time coal jobs are being replaced with short-term contracts. The new coal workers are younger, with no memory of the good old days. A few of them are at the Gundy pub. They work in the coal sector but identify as farmers.

Lifelong conservatives are advocating socialism, even if they don't recognise it. 'The government should be paying for everything,' they maintain. 'We help people in Asia, why not help people here?'

'I thought you wanted the government to leave us alone?' I say. 'To get on with our own business?'

The talk that night wasn't about the business of being on the land, but survival on it. And it's not just drought that's forcing change. Local farms have had to respond to global economic change. We're not Big Ag here, we're Small Ag, full-time and part-time operations feeding into a system where the real money is made way up the food chain. But globalisation isn't new, it's been with us ever since the colony was established. Change is a force in and of itself. And none of us know where it will take us.

All of us inherit knowledge and skills from the culture into which we're born. Big things like our religion. Little things like picking parsley. You can know a lot of things you know without knowing you do. For example, I don't know how I know the correct way to pick silverbeet, I just do. You choose your leaves and tear them off at the base, as you should with parsley. This stimulates regrowth. I also know how to turn silverbeet into spanakopita, the simple Greek pie in filo pastry. Blindfolded, without a recipe, without thinking. Making jam, sauerkraut or Christmas cake – all these are instinctive too. Daily rituals that help mental health.

Just as we inherit skills and traditions, so we inherit less tangible things from our forebears. It's been known for some time that cells use their DNA code depending on the task at hand. In the same way that a sports team knows the objective – to win – cells know the task they have to achieve. This is the case for every cell within us and for every cell in every living plant and animal and fungi. That makes for a lot of variation in cellular expression, and recently *Nature* magazine updated us on another force of nature: epigenetics.

New research suggests that experiences such as trauma may be carried or tagged to a gene and passed on for several generations.

For instance, if your grandmother was starved during a war, your body will carry innate knowledge of starvation. That is to say, we are modified by recent events, not just by the process of evolution. While our cells may remain the same, they behave a little differently. Genetic tagging may also provide some answers to understanding physical and mental health. Are we witnessing an increase in some health problems or an increase in the ability to recognise them?

This is fascinating because it captures a lot of what we feel about change on a personal level. Changes that are hard to comprehend and articulate. The science remains contentious and there are issues still to be addressed, such as, to what extent are we hostage to genetic tags with our own personal and cultural crises? Are these tags carrying guilt, shame and fear deep within our genetic cellular core? And do these tags just fall off after a generation or two?

At the other end of the scale, so many of the good things we've inherited can be easily lost in just a single generation – we can forget our parents' language, our mother's tongue. The cliché says that you don't know what you've lost until you've lost it; the reality is you usually don't even notice it's lost. Entire cultures can be lost without awareness or regret.

We know we're losing traditional agricultural skills all over the world. We're losing biodiversity in nature. We're seeing the loss of hundreds of languages, the loss of indigenous medicines that have been used for millennia. In Australia, song cycles have been lost, together with the secrets of fishing, the special skills and timing required for landscape burning, cave painting and kinship connections. The homogenisation of culture everywhere is making the world less interesting. We seek out unspoilt destinations and spoil them in the process. We fly in the same planes to identical airports with identical brands in the duty-free shops, and book into familiar chains of hotels. Soon there'll be little point in travelling. As in the episode of *Kath & Kim*, we might just as well holiday in an airport lounge and go shopping there.

Thanks to those profound discoveries in genetic science, we know that all humans of all colours from all times and everywhere are inextricably linked – surely a reason to abandon the nonsense of racial superiority and bigotry. Yet we're as reluctant to abandon this as we are to abandon our sense of superiority over the entirety of nature. This isn't the case only in Australia, but the effects are often more drastic here. This sunburnt country has proved hard to control and now much of it controls us. Damaged soil and empty dams and rivers lead to empty bank balances, diminishing rural populations and yes, rising suicide rates.

Our land at Elmswood is part of the Great Dividing Range. Old photos reveal treeless hills covered in prickly pear, and denuded and eroded paddocks devastated by rabbits and overgrazing. Now satellite images record some recovery. We've locked up 1200 hectares, to encourage nature to reclaim the valleys and the hillsides. Not everyone applauds this. Graziers express displeasure at what they see as 'old scrub' taking over, making it 'rubbish', 'useless for cattle'.

That triumphalism of the mythologised heroes who 'made

Australia' – the people who bullied the landscape into submission – continues today. It's the same arrogance that persists in land clearance, particularly that done illegally, by people indifferent to the law, in dirty forms of mining, in the live-sheep trade, in the stealing of water from the Murray–Darling, by people and companies who factor in the puny fines as a business expense, just like their political donations to the MPs who advocate for them in the parliaments. And it's there in the refusal to acknowledge that this has all been done on land taken from Aboriginal peoples. There are bullies aplenty in high positions where they ridicule climate change and inculcate hatred of dark-skinned foreigners, harking back to the days of the White Australia policy.

Millennia ago, what is now home to my garlic crop was a riverbed, its rocks first worn and rounded by the water, then ground into a sandy loam perfect for my bulbs. And while I'm continuing to change the environment and the ecology at Elmswood, I'm trying to do so in a way that builds soil as well as uses it. Give and take. Take and give. Of course, I cannot know the full consequences of my decisions, but new ways must start with new intentions.

Before this current drought, between 2015 and 2017, cattle prices rose to record heights. Some argued that this was just a catch-up after years of depressed markets, but either way, cattle farmers everywhere wanted to reap the benefits. Both high and low prices put pressure on a cattle herd. When everything is expensive it's costly building up a herd. When prices slump we're all eager to purchase more stock. In both cases farmers need grass and water in order to make money.

After a big muster in autumn 2017, we were at our cattle yards conducting a census and the count was down. Fifty cattle were missing. Lost, stolen or strayed? At approximately $1000 a head, this was a serious issue. We decided to send out a search party in the morning, but that night I phoned around the neighbours. There was always some coming and going with herds, always a fence down somewhere. More worryingly, there were always rumours of thieving, or cattle duffing.

Our newest neighbour is a giant mining company that bought a few thousand hectares to the west of us as an offset against their carbon dioxide emissions. Their property and ours share Black Mountain, which is high enough for snow in harsh winters. At the mountain's peak is the local trig point – a metallic marker that serves as a reference point for geographic measurements – alongside towers for the local fire brigade and Telstra. Our shared fence line would be classified by National Parks and Wildlife as a class 4 walk: it's off track and steep. I took an unfortunate friend with me on a brisk hike up there and she wanted to give up less than halfway. I wasn't only interested in the view, which is utterly magnificent, with our homestead's silver roof a tiny glint far below, I wanted to see how much of the meandering fence between the two properties was still standing. Some of it is eighty years old. It turned out that much of it was down.

Days after those cattle went missing, we found a few hiding in a protected valley near a half-empty dam. I bought a better night-vision camera online and moved it from place to place every few days. From a post near an all-but-dry dam (what was drinking there?) to a road (who was driving on it?) to the chook pen (what was stalking the chickens?).

Eventually, when the camera was placed in a tree above a back track, we found, along with images triggered by birds, kangaroos, feral pigs and wild dogs, some cattle heading in the wrong direction. They could have been just walking through a broken fence. Duffing themselves. They certainly looked like they were joining a neighbour's cattle count. Time to call in the Stock Squad of the New South Wales Police, although they're officially called the Rural Crime Prevention Team these days.

This led to months of investigation, on our behalf and on behalf of other neighbours who were missing animals too. Cattle identification, paddock movements, cattle numbers, cattle sold were all reviewed by a methodical police officer. No charges were laid but formalising the fear, making a complaint and not just repeating conversations with neighbours helped all of us reconcile the fact that we'd probably never see the cattle again. As the drought rolled on the duffers stopped duffing anyway. There was no grass and no water left for any cattle stolen or otherwise. But to be on the safe side, we decided to join the mining company in pulling down that eighty-year-old fence and replacing it with a mightily expensive, 100 per cent steel effort. A barrier worthy of Trump.

Good farmers know their landscape: the cattle tracks, the roo holes, the places cattle will head to on hot or windy days, the preferred watering holes and the times of the day they head there. A good stockperson may not know a paddock ecologically or be able to rattle off the botanical names of trees and grasses, but they can predict the behaviour of sheep and cattle. A study of the ground on a simple walk can tell them where and when the cattle have been, where they're drinking, what plants they're eating, even measure their metabolic function by the shape of their cowpats.

Everything about the way you choose to farm profoundly affects your whole local community, not just the neighbours sharing your fence lines. If you're a well-behaved occupant in a city apartment you try to make sure your plumbing doesn't leak, your noise doesn't disturb. But not everyone is so considerate in the bush. I heard evidence of this from an aggressive landowner years ago at a NSW Farmers' meeting, who told me, 'It's our land and we have a right to do what we want with it.' As if that gives licence to bulldoze bush illegally, steal water, fell timber, spray chemicals, scorn regulations. There are people everywhere who believe that any water pumped from a failing river or borehole is utterly, entirely theirs, that they have a right to keep their personal patch of land as an oasis when everyone else's is a desert. The animosity that these acts of defiance triggers can escalate into small-scale wars. The dobbing-in, the protests, the threats, the vandalising of pumps.

One recalcitrant in this last drought enraged us all by not

only pumping water from a borehole close to the river, but also spectacularly squandering it, spraying it on windy days. Literally pissing it into the wind. A daily downpour onto two tiny paddocks containing a grand total of two cows. It was tantamount to a golf course in the middle of the Nullarbor. I was getting calls from locals asking about their water use and whether we should ask them to stop irrigating. People knew that, as an active member of our now-defunct water group, I was intensely concerned about water policy.

Over a cup of tea I reflected on what I could say and how I might say it. 'I'm ringing you as a courtesy, not on behalf of anyone in particular, but to let you know that many people are unhappy about . . .' Or: 'Your ongoing irrigation in the face of the drought is causing a lot of concern.' Or perhaps: 'There's no river flow and using so much water for a patch of pasture for a couple of cows is, well, wasteful.'

When I dialled, the response to my attempt at diplomacy was abuse, first from the husband and then from the wife. 'People are envious!' they told me. 'We're just two pensioners! Ever since so-and-so moved to town there's been problems! Someone came one night and cut our pipe! They're vandals! We have a *licence!* We pay for it! IT'S OUR WATER!'

The put-upon pensioners kept irrigating.

There remains a total refusal to accept that by taking more than your share you're stealing from your neighbours. It's the Murray–Darling issue in miniature, a microcosmic version of the entire neo-con era of water management in Australia.

I confided in a friend about my unpleasant conversation. 'It's not up to you, Patrice,' my friend said sternly. 'The government should be dealing with it.'

When? I wondered. And how? There's not enough 'government' to keep watch over every single issue, to check compliance in every minor instance. Rural land management is up to us.

I'd feared that phone call would be a lost cause, otherwise I would've made the call weeks previously, but when I finally did make it I felt a little brave. But not for long. Weeks later I was snubbed by the couple at a local fair. I realised I'd broken an unwritten law, a law I loathe. In the country it's considered bolshie to openly criticise your neighbours, even in the privacy of a phone call. Behind their backs is within bounds, but never up front. It's one of the reasons social attitudes are so hard to change. Discussion of politics is held to be impolite. And in a similar vein, too many personal things are left unsaid.

Since that phone call I've reflected many times on how differently the conversation could have gone. Where precisely was the opening to mention something unpleasant? To share a new idea? What exactly is best left unsaid? Had I controlled my anger?

When Martin Luther King Jr said that we live in a 'network of mutuality', he didn't know that our genes were almost identical to those of the vegetables we eat. Genetic mapping has confirmed our interconnectedness with not just other primates, but with all of nature. But that little tiff over the health of our river and our dwindling water supply showed all too clearly how easily that interconnectedness can be torn apart.

CHAPTER 6

DIVERSIFYING

DROUGHTS HAVE A HABIT OF TRIGGERING A RETHINK ABOUT MY business. They're the best time to think hard about the land's potential, its capacity, and how to rebuild the farm once the rains come. When growth stops – the growth of your crops and of your bank account – that's the time to make decisions. And with Australia's greenhouse emissions soaring, making it unlikely that we'll not meet our Paris Agreement targets in 2030, I also need to consider, with every decision I make, whether it will help or hinder climate change mitigation.

As summer temperatures increase, it's uncertain whether garlic and olives will be feasible in the future. And with such high evaporation, secure clean water could limit the capacity to breed animals. Consequently, we're constantly adjusting the way we do and sell things.

Even before we started planting as many native trees as possible in strategic places on the farm, for shade, nectar, windbreaks and fodder, we already had dense tree stands, especially on our steep slopes. After years of thinking about incorporating agroforestry, the integration of farming with forestry, I lost personal and political enthusiasm. Dry spells, and nibbling and rooting animals, from hares to pigs, killed so many of our trees.

Trees are the aristocrats in a landscape, so the more we experiment with agroforestry the better, even though it's hard to see where and how financial gains can be achieved. For tree lovers, the best inspiration can be found in Rowan Reid's 2017 book *Heartwood: The Art and Science of Growing Trees for Conservation and Profit*, and from his non-profit Australian Agroforestry Foundation. Hundreds of people are learning from his example, supporting each other in the belief that chopping down a tree isn't the end of a local ecology, as long as we're planting them too.

Here at Elmswood we're yet to cut down a tree we've planted; increasing land value, biodiversity and shade has been motivation enough.

The National Forest Policy Statement recorded that in New South Wales, the public native forests managed for multiple uses were reduced from 2.6 million hectares in 1990 to 1.3 million hectares by 2008, due to their becoming conservation reserves. It's an ongoing point of contention that returning more forests to conservation could result in increased greenhouse gas (GHG) emissions in the long term. When the carbon dynamics and life cycle of wood are considered, forests managed for production over 200 years compared to conservation only have been shown to provide more GHG benefits and better long term carbon storage. The carbon cycle for different industries varies but GHG abatement could be lost if policy directives favour conservation over multiple-use forest production.

Many areas of Australia have been considered for forestry at one time or another. By 2008 the Forestry Corporation of New South Wales had identified 30,000 hectares of land around mining areas in the Upper Hunter as suitable for native timber plantations, even though the rainfall average of 640 millimetres per year isn't considered optimal. Perfect tree specimens may not be achieved but they can grow.

While I'd never consider spreading bio solids (from sewage) on the farm, their use on mine sites fits perfectly in the pursuit of building new soil after such momentous soil disturbance. And now native timber plantations on mine sites are being helped along with tonnes of poo. The new trees may not be replacing the carbon that was in the coal, but it is a small start. Coal was once plant matter.

While deforestation (the clearing of native or virgin forest, as opposed to plantations) is recognised as a major source of increased carbon in the atmosphere, there have been many studies assessing the difference in carbon uptake by young forests versus old forests. The carbon already stored in old forests outweighs the speed with which new forests absorb it. In other words, old forests provide greater ecological value than new forests. The report by the Intergovernmental Panel on Climate Change in 2000 reviewed many changing land-use options and concluded that the best way to sequester more carbon was to convert arable land to agroforestry. Forests *and* agriculture. We hear it again and again.

Calculations suggest that further carbon savings are achieved when forestry timber is used for long-lived products, as in homes and furniture. Leftover wood such as branches from thinning and post-harvest residues could be used for bioenergy. Forested areas on the farm are already biodiversity sites and carbon sinks.

All new business ideas such as agroforestry need to take into account not only the landscape, but also personal ambitions and way of life. Daily life on our farm begins for me with the morning ritual of opening the upstairs bedroom blinds, and stepping out onto the verandah to assess the night's activities. In a drought that usually means damage – the latest pig diggings, or huge kangaroos munching the last of the grass, some just below the window.

Perhaps a mist of rain, just a few millimetres, has created a hint of green on the shady sides of hills, a so-called 'green drought', which for a few days will look pretty. But it's not the short-lived shade of green that is important, it's the thousand shades of beige. The landscape's secrets are revealed during drought.

I walk around my dying garden remembering its promising past. The *Viburnum suspensum* is dead, rows of lettuces have been eaten, new growth of the *prunus* and *celtis* snipped off. A stab of green grabs my attention – the narcissus bulbs are out, triggered by the cold, as there's no moisture to help them.

Along the dry riverbed angophora droop gnarled limbs to the ground. They too feel the weight of the drought. Squire, our fourteen-year-old border collie–kelpie cross, trots beside me, keeping close. He can hear shots in the distance and has always been gun-shy. But then he becomes excited by something in one of the river's remaining waterholes, sludge holes now. I follow Squire. Is it an eel trapped and writhing? No, it's dozens of feral fish, accursed carp gasping for air in the muddy slurry. I consider manoeuvring the ATV down here, netting them out and making fish fertiliser. But it's too muddy, too hard.

With no compassion for carp I walk away, content that eagles or pigs will find and feed on them.

Brushing away the spider webs misting a tall stand of dead fennel, I head home, detouring to the desperately dry feijoa trees. I drag a hose over to provide a slow drip. Yesterday I ignored them. Today I want to save them. Feelings of hope come and go and come again.

Thirty years ago I planted an olive tree in the garden; it did well before the drought and still stands tall, if a little shabby. No slow drip for this tree. It can take it, whereas a tortured willow, despite receiving moisture from our septic transpiration, is almost dead.

To view the morning through a financial lens is to see problems everywhere. Green is the required colour for any agricultural imagery. All I can do now, in this brittle, brown-yellow world, is remember it. Green glorious green.

Still, it's comforting to know that the value of the farm, the core asset, remains intact, although diversifying is an ongoing process and always on my mind. There's an idea bubble most days, but in the way of bubbles, most don't make it to the next stage – in this case, of cautious investment and tentative experiment. To build a good farm takes a lot of love and care over many years. It also relies heavily, as we've seen, on community support, via the scores of ancillary businesses and skilled labour with which farmers engage. Transport companies to haul cattle and produce, shearers for the sheep, rural supply stores, fuel depots. Our regions must retain a critical mass of these things just to survive, never mind thrive. If the local saleyards close down, your transport costs soar. When fuel companies consolidate, deliveries become less frequent. And as regional centres decline they become less attractive to people raising a family. Before you know it, farmers can't get the help they need, not even a plumber or an electrician. You become helpless. Legend has it that rural people are resilient, but many surrender.

Community support also has an effect on the farm as an asset. In the rural sector, the concept of coexistence refers to one industry's right to carry out operations next to another, and in the Hunter Valley that means the mixture of coalmines and agriculture. But it also refers to farming right next to towns, which is not all that popular. Even if the farm long preceded encroaching homes, there'll be complaints about noise, smells and the spraying of chemicals.

The 'right to farm' idea developed in the USA in the 1960s, after Big Ag started building more and more feedlots, ending the bucolic ideal of animals happily grazing in picturesque paddocks. Who wanted to live next door to big smelly sheds? No one, as it turned out. To avoid being swamped by complaints, state after state introduced 'right to farm' laws, though the 'farms' protected by legislation were more like animal concentration camps.

In Australia, state laws overlap with federal regulations to cover land and water use. Here in New South Wales there is no overarching declaration of a right to farm. It's more a long list of environmental and industrial principles that should be adhered to. For example, not anyone can go to the saleyards and casually buy a cow. You need to be registered. If you receive a calf as a birthday gift, you can't just rear it in your back yard and then send it to market and collect the cash, unless you have a designated Property Identification Code (PIC). This PIC is included on every National Livestock Identification System tag. These special NLIS buttons are clipped into the ear of every animal so it can be tracked. From where it was born to where it was slaughtered, including all the places it may have lived in between. An NLIS tag is required for the sale of beef, sheep, goats, pigs, deer, bison, buffalo, alpacas and llamas.

This system ensures that all animals are registered to a specific property, and when every animal moves, that PIC and NLIS number moves with them. The Local Land Services keeps track of the properties currently registered to have animals. Should an outbreak of disease occur, the LLS can track the history of every animal affected. This is at the heart of our biosecurity, proving that we're a nation safe to trade with. It's like motor vehicle registration for meat.

I reflect on this when I ask Jane, a high-school friend of Aurora's who's helping us out before heading north to muster cattle, to please arrive very early, just before dawn, because before we clip the garlic we'll be butchering sheep – young lambs born on the property and not yet counted in our sheep ledger. There must be a lot of animals without numberplates in freezers around the nation.

We start butchering before the heat and, more importantly, the flies arrive. Last time, we experimented with cutting frozen lamb. Bad mistake. It broke the bandsaw blade. This time the lambs have been hanging in our new coolroom for two weeks, so should be perfect.

Although I sometimes think of becoming a vegetarian again, I still eat the meat we grow. The day I can't tolerate butchering will be the day I give up meat. Butchering helps remind me of the life of the animal and the effort that goes into putting meat on the plate. Meat is a very special and, yes, problematic, product.

Yet I enjoy butchering, particularly with Michael, who was Scone's best butcher until the arrival of Woolworths put him out of business. Michael also does the heavy lifting, heaving the half-carcasses on and off the hooks before quartering them. Jane and I aren't involved in the artistry of it – the fine cutting of chops, the deboning of a shoulder. Our role is to trim, bag and label.

Michael taught me most of what I know about butchering and all the different cuts of meat; we used to sell organic beef and lamb direct to posh restaurants in Sydney and to an Italian butcher who demanded the best. Now a coalminer, Michael keeps his hand in by butchering for me and a few friends. He makes it look so easy.

This is a barter economy. No money changes hands. We pay Michael in chops. And while we're working, I get to hear the latest from the coalface. The actual coalface, not the metaphorical one. Who's joined the mine, who's left, who's subcontracting. Whether the latest corporate takeover has made any difference to the management of the mine.

We're butchering Shropshires today. It wasn't so long ago that I travelled with Australian food writer Cherry Ripe to some inter-esting farms in her birthplace of Herefordshire. Small, charming farms with exotic pig breeds, unfamiliar sheep, and apple and pear orchards for boutique ciders. Some farms had biomass boilers to generate power and heat. (Despite pigs being such charming and highly intelligent animals, we've never kept domestic pigs at Elmswood, instead fighting a bitter and losing battle with their feral relatives, those four-legged bulldozers. One reason feral pigs

love our farm is that we don't graze animals beside the riverbeds. It's better for the soil and the water but it creates a pig heaven of soft grassy soil that's easy to dig up. Feral pigs are tough and fierce, thriving in even the driest of droughts. Like kangaroos, they hang out in mobs, support multigenerational family structures and love the nightlife.)

On one farm, Cherry directed my attention to some chunky Shropshire sheep wandering in the orchards. I learnt that this breed, thanks to a unique gut metabolism, isn't interested in tree bark. Bless them. Their preference was good old-fashioned grass. This wasn't the case with the black-faced Suffolks we'd had in the olive grove for years. Each year, around autumn, when winter grasses were germinating, the sheep turned to olive bark for nutrition. If we're not careful they'd ringbark the lot.

Back home I searched for a small mob of Shropshires to buy, and finally found a few in Victoria, under the care of founding director of the Rare Breeds Trust of Australia, Fiona Chambers. They weren't cheap and nor was the trucking, but they've been worth it. For three years now the Shroppies have left our olive trees unscathed and helped keep the grove tidy. They also nibble at the olive suckers that sprout from the trunks each spring and suck the life out of productive foliage.

Buzzing blade, sharp knife. Here's the fat around the kidneys, perfect for deep-frying potatoes. Here's the leg of lamb, the shanks, the best chops. I'm thinking recipes while we work. Here's the neck and flaps for dog meat. Michael cleans up by spreading flour over the bandsaw blade to absorb the fat and blood. We can never forget that meat comes from once-live beings. We raise them, kill them, cut them up and eat them.

Michael butchers six lambs. All the meat will be quickly eaten. That's what happens when hungry backpacking carnivores sit down and eat three chops each, with dogs at their feet waiting for the bones.

This morning, it's all over by nine. Cup of tea, toast and honey. Our honey. Then it's back to the garlic shed.

Farmers, whether they like it or not, are also in the real estate business. They've got one eye on the prices for their produce and another on the price of their land. As with the price of livestock, farm land values fluctuate, soaring during good times, sinking a little during drought. When farmers get old and the kids aren't enthusiastic about taking over, they'll look at the for-sale ads, with their pretty pictures of green paddocks, full dams, tidy sheds and homesteads with gardens, and wonder whether it's time to sell. Or is this the time to buy? While prices are down. Maybe make an offer on the next-door neighbour's river flats?

When thinking of selling, farmers can be shocked to learn what's valuable real estate. It's the raw land that's valuable, or not, depending on its position and access to water. Indeed, the property might be more valuable without extra pesky buildings; barns and sheds, cattle yards, garages, carports, even the beloved homestead and the swimming pool sometimes barely count. What good are a lot of silos if the prospective buyers are planning to grow vegetables? Visitors to Elmswood sometimes comment on the pressure there must be in having an old shearing shed. A building requiring maintenance isn't seen as an asset.

We've attended sad auctions for farm buildings – not only picturesque old sheds, but some huge modern ones. They'll be dismantled, carted off and rebuilt somewhere else. It's good that there's recycling, but all too often buildings are bulldozed simply to get them out of the way. When I pass derelict wooden sheds with quality galvo roofs I yearn to rebuild them at Elmswood.

So, how much to spend on asset building? In 2015 we started a long overdue series of investments in infrastructure with the building of new cattle yards. An essential item for the beef operator, these yards are not cheap. The basics begin with a crush, a holding area where an animal can be safely positioned and checked over and the PIC–NLIS tag fitted in its ear. The crush is also where a cow is tested for pregnancy, and the scrotums of bulls are examined to ensure they're in good working order.

Crushes come in lots of shapes and sizes, some basic and others fancy. We saw one we liked at Agquip, a huge agricultural show at Gunnedah, signed on the dotted line and drove home excited. Within days we were embroiled in arguments. Here's a sample of the one-sided correspondence:

> When we made a long trip to visit the Red River display at Agquip we were impressed. Your staff seemed competent, enthusiastic and professional. So much so that I broke a lifelong rule regarding prepayment and paid you full price in advance for a cattle crush/weighing system. As you well know, I've had cause to regret this.
>
> You received my payment in August 2015. It is now February 2016 and your equipment still fails to operate.
>
> Firstly, it arrived three months late. Then we made the painful and expensive discovery that it didn't work. It still doesn't. I'm sure you're aware of the consumer protection laws that apply – and of both the legal and ethical issues that arise.
>
> It is not the usual practice for a manufacturer to complete his R&D on the customer's premises on their dime.

Consider the additional labour time for my manager and staff, the vast amount of wasted time and stress for me, and the fact that you've effectively put our cattle business on hold . . .

The response? Utter silence. The lesson? Never pay in advance. We wanted to put Mr Red River in his own crush and examine his testicles.

But a crush and weighing system is only part of the investment of yards. Months later, with most of our cattle sold because of the drought, the yards slipped down the priority list. Even quality yards aren't a saleable asset on a farm. Every set of yards is unique, just as its position is, and every stockperson I've met has strong opinions about how their preferred yards should operate. I often admire what seems the gold standard, only to have someone rubbish them.

After our cattle crush woes subsided, it was time to buy some other unaffordable things we'd wanted for years. Garlic being so labour-intensive, us labourers needed more mechanical help. I love old things, the reinvented, the recycled, but sometimes new gear is essential.

Our garlic business had started without any investment in equipment. The most important tools during year one were our hands. Graeme, though, loved ancient farm machinery, especially the Massey Ferguson 35, a tractor as old as me. So after that first crop he looked all over for one. Sometimes we'd drive past a farm and spot one rusting in a shed. We'd call in – was it for sale?

But most Massey owners love their oldies. They were part of the family, of the farm's history, and those we asked were always planning to do it up sometime.

Imagine Graeme's delight when we visited a garlic grower at Waikerie, on the Murray River, and got a guided tour of his treasured tractors. But he too had none for sale. Eventually we did find the perfect Massey, persuaded the reluctant owner to sell it, and had it delivered to Elmswood. The red paint had long since faded, the tyres were flat and the engine needed work. Graeme liked it that way. It meant he could do it up.

Every true farmer loves their old gear and tries to keep it running. That's why you see on every dinkum farm what I call Heath-Robinsonian machines, because they're made up of all sorts of spare parts. Over the generations, farms become mechanical museums, with most farmers also being good bush mechanics. Reg was, then Phil, who took over from him as manager, and Graeme and John. Each one of them would stare at a broken machine for a while and then get stuck into it. They could read cogs, levers and pulleys like a musician reads a score, and soon the machine would be alive again, digging post holes, slashing weeds or baling hay. And our managers were the same with their own cars and trucks, most of which were veteran vehicles if not technically vintage.

The new machine I most wanted had a fine name: rehabilitator. This is a kind of one-pass plough that can go through a good stand of grass and break it all down. Pretty soon you have a bed ready to plant. I also wanted a weeder. The organic sector has a few different kinds and many are specific to particular crops. Then we'd need a planting machine and a harvester. And to have all the new gear work successfully we needed a different tractor, the old Massey after eight years of loyal work wasn't big enough or modern enough to handle the equipment.

Our biggest tractor (built by the Lamborghini company in Italy, a 'Lambo') was being used for ploughing and heavy lifting, but is not a horticultural tractor, which needs both a degree of delicacy and narrow tyres. So at long last we bought one. A Deutz, which could take all the new gear up and down the rows. This lot was hugely expensive in the short term, but in a few years, I reasoned, we'd be ahead.

Our unfinished cattle yards don't bother me as much anymore, but I have been upset about the slow death of the narrow-leafed ash trees, *Fraxinus augustifolia*, that we planted thirty years ago to shade us and the cattle. After being helped through their early life with buckets of water and tree guards, they grew strong, providing shade through many summers, despite horses chewing their bark off and cattle nipping the lower limbs. Now with gnarled trucks and wide canopies, they've suffered during this latest drought. They need a nursing home, but I can't move them.

It isn't only the cost of infrastructure that can be lost when it comes time to sell a farm; the time and effort and cost devoted to farming organically can also go unrecouped. Unless a prospective purchaser shares your ideals, it doesn't much matter if your farm has been drenched in chemicals or not. The buyers may be tree-changers who fall in love with the house, the view or a bubbling creek, in which case there's no bonus for having tended the land carefully for decades, for rebuilding the soil, for investing in different ways of thinking and being.

Any farm investment is a risk; a single weather or market event can undo a year's work. As well as calculating the size of the loan we'd need for the new equipment, I needed to make calculations about my life span and health. But few emotions compare to the excitement of growing something, and for this reason I decided that garlic was worth another ten years of my best efforts. I love garlic the way Graeme loved tractors – I loved every part of the process, from when we put a clove in the ground until we put the bulbs into boxes for a customer nine months later. So we bought the tractors and other marvellous machines. Happy birthday to me.

Big questions about new infrastructure and succession aside, there are the daily duties on the farm to do. Our former manager, Gavin, returns to help with the garlic harvest. Back in 2006 he drove out the farm gate, after working here for five years, with a massive catamaran mast he'd built. He has since been living the sailor's life travelling around the Solomon Islands. I hand him an endless list of other tasks as well as the garlic: repairing and replacing some of the rotting footings in our classic shearing shed, and temporarily converting more of it into a bigger garlic-drying shed.

With the agility of a lifelong sailor, Gavin slides beneath the shed to check on the footings. Most of the wood is from the original, late-nineteenth century construction, preserved by the lanolin that oozes from the wool. The floor is almost wavelike in sections, and termites have done their worst, chomping through posts. Dripping water has caused more damage, thanks to the unrepaired roof. Thousands of sheep have donated their droppings over the decades, and as Gavin crawls around underneath he calls out to ask if I want some for the vegie garden.

As I worry about snakes down there his head suddenly appears through a hole in the floor. 'No snakes!' he grins, reading my mind.

Later, he and Betty weave some number-8 wire and strain it from beam to beam in the shed. When we hang the garlic bunches we can hear the old timbers creak like a sailing ship's, so we move from fore to aft and wire up another area to share the weight. Below the wires we arrange crates and poles to load more and more bunches. Shearing sheds are designed to catch airflow, to cool the hot task of shearing, and this makes them ideal for drying garlic. There only needs to be the smallest breeze and you can feel it breathing over the garlic.

If the crop is to reach its full garlicky potential and maximum value, each bulb needs to be generous in size and cured so that mould doesn't take hold. The moisture in each bulb makes it heavy, so a two-tonne crop, once dried, becomes a one-tonne crop.

When we plant garlic close to the public road that runs through the property we can see our neighbours drive by, following predictable, daily timetables. One regular, every morning, is Kate, who unfailingly gives a toot and a wave as she drives out. And we all look up from our work to smile and wave back. Kate helped pack garlic over the years and has ridden her horse across the paddocks when our stock are mixed up or go missing.

Never again. In 2012 there's a head-on accident on the Gundy road, where it winds towards Scone. Hours later it is confirmed that Kate is dead. The news ricochets around the district. Kate was the new wife of our dear neighbour Des, the quietest, shyest, most decent of men. Until that morning they were two of the happiest people imaginable.

For weeks we talk about the day Kate drove by and gave her final toot, her last wave. Years later a coronial inquiry makes it official: a mechanical fault caused her vehicle to swerve across the double lines. To this day, every time I drive past the site I think of beautiful Kate.

Vale Kate.

When I don't get to read a paper or watch TV, I do a lot of walking up and down the garlic rows with my iPhone in my pocket, headphones in my ears, trying to keep up with the news. Late in 2017 I heard one of a swathe of so-called experts saying we were about to become a borderless world with genderless individuals. Some WMABs – white middle-aged blokes – were coming to the bitter realisation that the patriarchy is running out of puff. Thank heavens.

When Malcolm Turnbull, then Prime Minister, finally launched the Marriage Law Postal Survey, most of my friends

were already over it. Some are gay, all have a gay friend or two. For Phillip and me the topic was prehistoric. We'd already had a same-sex marriage ceremony here at the farm. Not technically legal, but a great occasion. Aurora fell in love with Susannah when they were studying at Edinburgh University. Their ceremony took place in the hot outdoors, with flowers and hay-bale seating, witnessed by friends from around the world. It was followed by all-night dancing in the garlic shed.

On 15 November 2017, I leave the garlic and go inside to watch the result on TV. Nearly 13 million Australians have voted, meaning four out of five eligible people chose to participate. The numbers are announced. The Yes vote gets 7,817,247 – a comforting 61.1 per cent. I feel a cloud lift knowing my daughter and Susannah are in a better country.

A few months later, I received a text message announcing they'd attended the registry office with a few friends. Now it was 100 per cent official, complete with forms and photos. Oddly enough, there were few in-focus photos of the Elmswood event. Too much dancing in the garlic shed.

Without commercial productivity it will be unaffordable to stay here. Rates, taxes, levies, insurance, maintenance – all cost money. My research into product options is ongoing, not just in terms of what we could grow in a changing climate, but also how to grow it.

By the end of 2018 the consequences of the drought meant

that we were left with a garlic crop a quarter the volume we'd normally harvest. Not just fewer bulbs but each bulb was smaller. And we had no olive harvest at all, because there'd been no water for irrigation. All our dams eventually dried up, and the water used for the trough system could only be pumped for two hours a day as the river's last waterhole was diminished. We'd gone from having many hundreds of cattle to just fourteen, and we only had those because they were missed in the final muster. Our two mobs of sheep were attacked by wild dogs, lambs were lost, and soon we were feeding the remainder hay to keep them alive. She-oaks that had reclaimed the riverbanks after we fenced them off from stock were dying in clusters. We hadn't seen a platypus in years. Short-necked turtles died as the river retreated, their carapaces littered across the dry riverbed. We collected one, then two dead turtles, then John returned with a bucketful of them, a bucketful of sadness. Their shells a miracle of design.

Every other year, we'd had to stop and pick up turtles that were about to cross the road, so that the next car coming along wouldn't crush them. Some drivers actually swerve to hit them. Now wild dogs are suspected of digging up turtle eggs and wiping out whole families, and we fear these ancient creatures might become extinct.

The native *Angophera floribunda*, often known as the rough-barked apple but also called Christmas trees around here, starts its prolific summer show of flowers by 25 December, and in a normal year I watch our beehives fill will its dark nectar. But during the past two years the flowering has coincided with days of 40-degree heat and less nectar is making it into the hives. No longer do we have a guaranteed supply of summer sweetness.

This is the lesson we're learning incrementally. What part of nature can we rely on? Will the nectar flow for honey? Will the river flow for garlic or olives or turtles? Will the grass grow for sheep and wallabies? Alongside this, no matter how disquieting, is the hope that behind the failure to make money during a drought we're surrounded by nature, appearing and disappearing, living and dying. Eternally flowing.

CHAPTER 7

REGENERATION

IT'S GOOD TO HEAR A FELLOW FAMER SAY, 'I'M A SOIL REGENERA-
tor, really,' in a matter-of-fact way. But it's not only the soil that
needs regeneration, nor is it just agriculture, it's us. Agriculture,
in my view, should be aligned with the health sector, not merely
to resource management.

Agricultural production, whether it be food, fibre or fuel, is
not the only concern. If someone wants to buy land and use it for
fun, what difference does it make so long as the land is regener-
ated? This idea upsets those who see land use as being based on
increasing exports, a percentage of GDP, or pushing up agriculture's cash value. Placing agriculture with healthcare would give
us an association beyond money. Our health is our most prized
personal possession.

Going into garlic production meant that I'd officially entered the retail sector, but land management occupied more time. And while I classify Elmswood's agricultural operations in a personal and idiosyncratic way, they must also be rendered official, by filling out economic activity surveys for the Australian Bureau of Statistics. I must declare the number of olive trees, the area given over to crops, the area of native vegetation, and the number of domesticated animals. Most of the farm's land is classed as broadacre management. The river flats, with the garlic patch, are classed as horticulture; our six thousand olive trees are classed as fruit production. And all of the property is deemed apicultural.

At a local private field day in 2018 I listened to a first-class farmer, now a consultant, talk about natural fertilisers. The drought had been drawing ever more desperate farmers to such field days, and here there were more than fifty, wanting to discuss photosynthesis and soil requirements. The farmer/consultant had a reputation for straight talking, and now he was trying to start a critical conversation about climate change with a group of people who mostly think, and often say, that it's all a load of hogwash.

Our plain speaker began with a compliment, saying that all farmers are the great carers of the land, but over the next four hours the emphasis slowly changed from the congratulatory to our litany of past mistakes. The land has been degraded, biodiversity lost, rivers polluted, soil compacted and eroded all causing a serious loss of soil function and soil itself. The plain talker turned tough but judging by the nodding heads he was also therapeutic.

All of us are managing land that needs repair work. When we purchased our property the real estate agent didn't point out sites of erosion, gouged creeks, soil compaction or weeds. He was selling us productive land that could be made even more productive with fertiliser. The value of the land was based on its potential carrying capacity, not its nature. No past environmental damage was factored into the price, just as the further damage being done in this drought is unlikely to impact a future price. Rural real estate should be in freefall, yet it isn't. In our collective psyche we must have faith in the principles of regeneration. But if maintaining biodiversity during a drought was a law, most of us would be in jail.

As of 30 June 2017, there were 394 million hectares of land deemed agricultural in Australia, a 6 per cent increase from the previous year. There were also 2400 more agricultural businesses, representing a 3 per cent increase since 2015–16. But there's not been a 3 per cent increase in agricultural employment;

that continues to decrease. The value of all primary production was $61 billion, less than 3 per cent of GDP, and the percentage is falling along with farm employment.

Not long after I became a novice farmer, New Zealand feminist, economist and MP Marilyn Waring wrote *Counting for Nothing*. The book became my New Testament. In it Waring argued that GDP as currently calculated was dangerous nonsense, because so much work done by women and children and by people living outside the cash economy was unaccounted. (Gender equality is one of the UN's sustainability goals yet figures in 2018 show that rural women have a long way to go with women only owning 13 per cent of agricultural land globally.) Waring has spent a lifetime arguing with precision and passion about the lunacy of an economic accounting system that doesn't account properly for nature, but at last is being discussed at the highest level of financial management, academia and politics.

After the 2006 Stern report on the economic consequences of climate change, the idea of putting a true price on global environmental damage gained some traction. On farms this would include the damage caused by the misuse of ploughs, soil compaction, fossil fuel used in transport and fertilisers, chemical use and land clearing. In 2010 another report analysed thousands of the world's biggest companies, not just agricultural ones, and estimated that a third of their profits could be lost were they required to pay for the environmental damage they've caused. Environmental problems are also climate change problems. We now have the UN's Principles of Responsible Investment, a guide to help companies everywhere take up new fiduciary duties to slow climate change and meet global obligations.

Exporting 70 per cent of our agricultural production is important to Australia's balance of trade. If we're to continue to export at this rate, we must have some means of distinguishing positive actions from negative actions on farms. At present, regenerating land is usually seen as pure expense. We need it to be profitable. As Marilyn Waring points out, when wars destroy things – roads, bridges, entire cities – there's money to be made in repairing the ruins, with the added benefit of replacing the old infrastructure with new. Bombs lead to booms. Look at postwar Germany and Japan.

We're seeing some recognition of this in the growing number of conferences on erosion control and waste, especially in the mining sector. Engineering solutions and products to repair damaged landscapes abound. Farmers are small-scale engineers themselves, redesigning machinery, adding extra tyres to ploughs. At Elmswood we welded our irrigation tape layer onto our mechanical weeder so one machine did two tasks. Our new garlic harvester is always being adjusted to tie different-sized bunches.

There's a yeoman in Chaucer's *Canterbury Tales* and Robin Hood was another. Originally the term was linked to nobility, later evolving into 'yeoman farmers', men of property. Australian history can boast a wholly inspirational yeoman of its own – the wonderfully named Percival Alfred Yeomans, an engineer and farmer in the North Richmond district of the Hawkesbury catchment in New South Wales. During the late 1940s Yeomans began developing what is known as the Keyline system of land management, focusing on land contours to plan farm operations. Using a design of his own invention, the 'Yeomans' plough', he found a better way to aerate soil. The techniques are still taught, seventy years on, in many university land management courses and are incorporated into permaculture design principles.

Another farm implement, the Agroplow also helps. When we first bought the farm it was on a 'walk-in walk-out' basis, which meant we got the stock, the gear, the lot. We found an almost new Paraplough in the shed – a third version of the Yeomans and Agroplow, also designed to open up soil and guide water movement.

Another yeoman visionary joins the narrative this century: Peter Andrews came to national prominence via the ABC's *Australian Story*. Peter's clarion call is that water movement through soil can be fine-tuned when the shape of the landscape is properly understood and plants allowed to thrive, thereby allowing more moisture to be held in the soil. Peter urges land holders to copy the natural processes that once sustained us, and to never forget the role of gravity. ('A contour line and gravity will define everything' he once told me). Peter is a lover of plants – ALL plants including weeds – proclaiming their importance in healing the land. Plants build landscape function via photosynthesis, so the more plants using the sun's abundant light the better the landscape will be.

Peter recommends where possible a 'leaky weir' system, literally weirs that leak, to slow down the movement of water in gullies and streams. After decades of experimenting, Peter's techniques have shown splendid results. Though many of his bigger projects have been funded by government grants, paradoxically some of Peter's ideas are still contrary to land-use laws.

'I'm a messenger,' says Peter, 'trying to help people understand our landscape.' Now in his late seventies he's been working on his latest property near Goulburn in New South Wales reinstating landscape function with the power of plants, sunlight and gravity.

When dollars rather than decency define success, it's a challenge to incorporate regeneration into our personal, professional and political lives. Peter paid a high price when he started sharing his profound ideas, losing his farm as he focused on developing systems in order to save other people's farms, the classic case of the prophet without profit. The very word 'regeneration' helps us imagine a cyclical rather than linear economy. Let a new generation use regeneration to defeat the brainwashed belief in never-ending economic growth, or at the very least push this malignancy into remission. Cells can regenerate. We can regenerate. So can ailing societies and ailing land.

Economists are also beginning to address the value of unpaid work done by women and children. A 2018 British study put a price of £1.2 trillion a year (about A$1.6 trillion) on this. Driving children around represented the largest slice of unpaid household work, at £358 billion, followed by childcare, at £352 billion. Cooking – referred to as nutritional services – came next at £158 billion (I hope we don't start saying, 'I've got to get home to provide our nutritional service'); and clothes washing at £89 billion.

The future work of repairing land shouldn't just fall to environmental enthusiasts, rich tree-changers or Landcare Australia members. It must be shared by everyone who owns or leases land, and that includes farmers: those living and working on the 80,000 operations recognised by the government as family-farm businesses. And given that on a family farm it's often the whole family working to keep things going, there's already an in situ green army. The family farm, no matter what size, should be reimagined and given wider compulsory tasks in order to receive government grants, such as drought relief and diesel rebates.

Otherwise the future will mean increasing dominance of corporate agriculture, and there is not a scintilla of evidence that corporate farms are better environmental stewards.

History has shown, and Australia's highly successful Landcare movement confirms, that local communities are often the best at land regeneration, assessing, monitoring and conserving biodiversity on all types of lands, both private and public. Google Maps and satellite images are good, but people on the ground are better. Even suburban home gardens are important components in preserving biodiversity and potential providers of biomass.

When Telstra began to fully privatise in the 1990s a lot of money was generated for the government. A billion dollars was used to set up the National Heritage Trust, which funnelled hundreds of thousands of dollars into our area to repair erosion, build biodiversity and fence off rivers. That money was available because a precious asset had been sold: with diminishing public assets left to sell, other means, like taxation and levies, will need to be found to fund future environmental repair work. Those areas that were brutally cleared with government funding now need to be reforested with government money.

At the time of the Telstra sale, rural areas were also promised $2 billion to attack the black spots in our communication networks. Where these allocated funds went is anyone's guess, because years down the track the spots are even bigger and blacker. And it's not just marginal land where farms are off the grid either; in our area some of the best farms struggle with communication infrastructure.

The UN's Food and Agriculture Organization suggests that the term 'marginal' should not be based purely on a land classification alone, because marginality can be modified by on-site investments and factors affecting markets, prices, resource entitlements, etc. Marginality of land can be human-induced. Land can move quickly into and out of marginal status depending on many factors but money helps. Farmers often manage both marginal and fertile land on their property.

Recognising the need to integrate farms within communities is important – to take advantage of the local resources such as labour, technology and surrounding infrastructure. Location relative to infrastructure such as roads, railroads, harbours and cities alters economic returns. What is deemed marginal for one area might be valued in another.

Australia's agricultural development has historically taken a broadacre industrial approach. It relies on a diminishing selection of new plant varieties, the manufacture of synthetic fertilisers and pesticides, increased farm sizes and the introduction of large-scale mechanical equipment, an approach deemed innovative as new technology and allegedly science-based, yet that continues to degrade ecosystems. It's a system that has increased the distance food and fibre travel to the consumer. The system has led to a reduction in soil carbon. Industrial agriculture is different from a regenerative and sustainable agricultural model that advocates the cyclical use of natural resources, does not rely on synthetic fertilisers, can be profitable at a smaller scale, promotes local distribution first and aims to sequester soil carbon.

The assumptions about economies of scale in industrial agriculture are highly questionable. Large farms can actually use a higher percentage of energy per unit of production than smaller ones. Some estimates have industrial agriculture consuming fifty times the energy of small, mixed operations, which can set the example in the sequestration of carbon, and can be productive with fewer input costs. There are also cultural and ecological benefits with smaller, more traditional farms, when they are managed properly, yet all this is overlooked by politicians and agricultural planners in favour of big-farm policy.

Back in 1993 the OECD claimed that the absolute energy consumption per hectare of agricultural land globally had increased by 39 per cent between 1970 and 1989. And since then, fossil fuel inputs, directly and indirectly, have led to agriculture becoming even less energy-efficient. Australian Bureau of Statistics figures for the period 2011–12 reveal that fertiliser was applied to 46.7 million hectares of agricultural land across Australia, an increase of 6 per cent from 2009–10.[17] Increased use of synthetic fertiliser results in increased climate change emissions. The challenge is to build soil fertility without fossil fuel inputs. Farming systems that integrate trees with cropping and grazing have shown an accumulation in soils up to 0.3–0.6 tonnes of carbon per hectare. Those that include native wildlife, such as kangaroos and wallabies, as alternative grazing animals to livestock may also help reduce methane emissions. Our native soil-turning animals, such as bandicoots, bettongs and potoroos are also working as unpaid ecosystem engineers across many farms as they gently till topsoil, spreading seeds and leaf matter.

The negative impact of poor land use was highlighted at the world's first conference on biofortification in 2010. Biofortification is the breeding of crops for improved nutrition, as opposed to adding minerals and vitamins during processing. Micronutrients, the conference confirmed, are now reduced across most soil types, a consequence of the depletion of soil rather than its development. Ensuring that micronutrients are not further reduced is vital for human nutrition: if the nutrients aren't in the soil, they're not in the crops.

When scientists conducted qualitative research with graziers who were changing their land management practices it showed that prior to their Holistic Management training, only 9 per cent of the respondents considered biodiversity in their business decisions. Afterwards all of them did. These small steps are part of the solution to repair the mismanaged natural resources everywhere that Responsible Investment PRI and the UN's Finance Initiative estimate cost the world economy US$6.6 trillion a year.

Much credit should go to the inspirational work of Zimbabwean farmer and scientist Allan Savory who designed the principles of Holistic Management in the 1960s. Through observing wild animal herd behaviour and how it impacts plant growth he redefined how we should be looking after our domesticated animals and plants.

While every farm will require unique strategies, resting land from livestock rather than completely destocking has proven to be important in building grassland resilience. This varies depending on the level of historic destruction. Farmers around the world are now applying it. We use special holistic managing charts to record grass growth and all stock movements across the farm. These record what we plan to do, depending on the grass availability and then what we actually do. They are also the best tool in helping to decide when it's time to destock. Keeping grass cover at all times is essential.

The alternative to Savory's idea is called set stocking, where animals are left in the same paddock eating the same old diet, often not letting plants recover from being chewed and all the while compacting the ground. Another response to previous grazing errors is to remove domestic livestock all together. This idea has been a key component in government-endorsed Property Vegetation Plans. However, removal of domesticated animals 100 per cent of the time can also have unexpected negative ecological impacts such as reducing biodiversity. I witnessed this at the farm firsthand, when after 15 years of not grazing in a tree lot, our local Holistic Management group, during a field day investigated how the grasses were going. It turned out the pasture was in a worse condition than the adjacent area where we rotationally graze cows and calves. It was one of those aha! moments. I'd been looking at the trees, not the ground. There were areas of bare soil around tall dry clumps of grass and because they were never chewed, they shaded out the surrounding area. Now that the trees are bigger we let stock in from time to time to keep the grasses under control.

Rotational or strategic grazing is now the preferred practice, recommended by New South Wales Local Lands Services and promoted via animal management education programs. Savory suggests this is a prime method to halt desertification.[18]

With so much of eastern Australia in drought from 2017 and into 2019, you might think that beef statistics would show decline. Yet there's still almost 27 million head walking around our wide brown land. That's 2 million more cattle than *Homo sapiens*.

Many environmentalists advocate meat-free days or vegetarianism, but anti-meat arguments should really focus on industrial meat, the production of which is unhealthy and inhumane. Chickens and pigs are also being raised in cruel factory-farm feedlots, which are infamously a source of massive soil and water pollution.[19] Around the world, victories against these animal concentration camps are on the rise. A North Carolina jury in 2018 awarded more than $50 million in damages to the neighbours of a 15,000-head pig operation – not because of the inhumane treatment of the animals, but because of the impact of the smell, spraying, noise and pollution on humans. The payout was later reduced due to a state law capping punitive damages. It was a small loss for the company's bottom line and failed to address the core issue of keeping 15,000 animals in overcrowded sheds.

It's irrefutable that animal production creates GHG emissions, in the form of methane, from enteric fermentation (during the digestive process) and manure.[20] This represents around 18 per cent of total global GHG emissions. This coincides with the global increase in meat consumption and the increase in Australia's meat exports. But we shouldn't forget that humans emit gases too.

Grassland animal husbandry in fact has the potential to concurrently reduce GHG emissions and increase production. Animals are essential to ecology, and they can actually build soil, through their manure. They also stimulate grass growth as they graze; they're technically pruning, which increases plant vigour, just as the pruning of trees and vines by humans stimulates fruit. Domesticated and native animals grazing a paddock can therefore build grass volume, and that in turn sequesters carbon.

Anti-meat advocacy often ignores these facts, so let me say loud and clear that not all meat eating is bad and not all meat production is bad. What's bad is meat production via feedlots, where cattle eat grains instead of grasses and are forced to live miserable, unnatural lives trapped in their own faeces. Nor does a vegetarian or vegan diet avoid the issue of killing for food. Grain crops have killed off countless local ecologies that once supported native birds and animals; they've also resulted in the killing of insects and soil microbes, even more so when crops are sprayed with chemicals. Unless non–meat eaters manage to be 100 per cent organic, they're complicit in the chemical warfare against multiple forms of life. The act of growing any kind of food demands sacrifice.

Neither animals nor plants can argue their own case, of course, but there's now evidence that plants are societal and can actually communicate, perhaps even have some sort of feelings.[21] We're all part of an ecology, living and dying together.

Livestock can have an even more important role to play when biochar is brought into the picture. Experiments with charcoal in the diet of ruminants have found that char can assist digestion. Grazing animals that eat char could cost-effectively integrate carbon sequestration while producing meat. The absorption qualities of char have also been found to bind toxic substances in the gastrointestinal tracts of cattle.

Powdery char is easier to handle as a feedstock than as a soil additive. Its health benefits are delivered simultaneously to the stock and the soil, via faeces and dung beetles. There's no need to mechanically prepare the soil or consider complex crop nutrient requirements, nor the use of a tractor when spreading char across a paddock. Char placed in a feeder for animal access is sufficient.

Char has been trialled specifically to improve animal health and decrease methane emissions. Two experiments during 2012 found that adding as little as 1 per cent of char to feed reduced methane emissions by 12 per cent. Further testing, whereby char was mixed with cassava (not a common cattle feed in Australia, but it is elsewhere), showed both methane reduction and weight gain, thus confirming that char can improve feed conversion in the rumen (the first stomach of ruminants).

In Japan, char remains a common medicine for animal intestinal disorders. The highlight of my first biochar conference in 2009 was hearing from Japanese farmers feeding their chickens char to improve their health. And when pyrolysis liquid was mixed with the char, it seemed to improve animal health even further. An experiment published in 2014 showed that when 200–400 grams of char was fed to a cow each day, its health and vigour improved. And as a bonus, bad odours were reduced. But even if cattle are given as little as 300 grams a day, significant quantities of char can be spread across a paddock, easily and slowly.

The soil carbon in Amazonian soils didn't just get put there once. It was developed slowly over centuries and revealed in the layering in the pits I explored. It is unlikely that mechanically applying char just once to soil will achieve the same results with carbon sequestration. Feeding char to the soil via the gastro-intestinal tracts of ruminants – and allowing earthworms to assist in the spread and penetration – seems more effective and practical.[22]

In 2015 there were 62,811 head of cattle in our postcode area. If all cattle across Australia were to consume small volumes of char regularly (300 grams per day), this would allow thousands of hectares of grassland to sequester far more carbon. Even if we all became vegetarians en masse, we could still use these existing animals as gentle helpers to develop soil biology and sequester carbon. When it comes to mitigating climate change, soil carbon is important. I'd like to see biochar as part of our future national soil development. Whether they know it or not, farmers are forever moving carbon around anyway, sometimes doing the right thing by accident, or by benign neglect. Whenever they don't destroy native pasture or cut down trees or burn the old ones that died decades before from natural causes, they're doing the right thing.

On a bad day I'll notice more the local political and social conservatives who view climate change not just as a scientific fraud, but sometimes a communist homosexual same-sex sharia-law Muslim plot. And I'm only half joking. Rural Neros fiddling while Rome burns. Recently a local councillor asked me to stop talking about climate change and start saying changing climate. The name-changers seek to not only confuse but to muffle the debate.

We all have to learn to tread more lightly on the land. Even gentle tilling releases carbon from the earth. We know that soil stores carbon, we know how to increase it, and we're learning to measure it, to take it into account in the endless arguments about emissions.[23] As much as solar, wind, tidal, geothermal and other non-fossil fuels, it's soil that matters. The earth can literally save the Earth.

The issue is how. There's always been a disconnection between the agricultural business that overlies environmental management – or, more accurately in most cases, the lack of environmental management – and the act of farming. In the best of all possible worlds we'd demand that businesses avoid collateral damage.

There are plenty of good ideas out there about how things can be done better, and plenty of good books, telling different parts of the same story. A favourite of mine is *Soil and Civilization* by Edward Hyams, published in 1952. In a calm, thoughtful and slightly peeved voice, Hyams gives a potted history of how different cultures have treated soil and the extent to which they have succeeded or failed in its care. There are good soils across all different geographies, but the locals haven't always been smart in managing them. Greedy humans have mistreated them, wrought havoc, created deserts. In a few places but specifically Peru, Hyams was impressed how humans thought of themselves as being at one with soil, becoming soil makers rather than destroyers. Bruce Pascoe's 2014 book *Dark Emu* describes numerous Indigenous agricultural systems across Australia where deep, fertile soils were managed for millenia before the first European settlers arrived and destroyed them.[24] In the twentieth century, further damage was done on land allocated, after both world wars, across South Australia, Victoria, Queensland and New South Wales under the Soldier Settlement scheme. Many of the soldiers who farmed this land were ill-advised and doomed to fail. The *Phosphate Fertilizers Subsidy Act 1963–71* resulted in large volumes of superphosphate being applied to improve yields in the short term, but this did nothing to build soil fertility in the long term. Numerous publicly funded irrigation schemes supplied the water to liquefy the fertilisers.

In Europe, the Mansholt Plan of 1968 redefined agriculture with its ambitious agenda to ensure affordable food. It encouraged bigger farms to operate more efficiently. The plan predicted that 5 million hectares of European agricultural land would, and should, be removed from production. Not surprisingly, this revolutionary restructure was not met with enthusiasm by farmers. It was the beginning of agriculture being viewed politically as just another business, without recognising the unique ecosystem services and biodiversity interconnections that were also involved.

The EU's Common Agricultural Policy now requires all farmers to have unfarmed 'ecological focus areas' to protect soil and for nutrient retention. Crops are planned and registered to ensure diversity is maintained. Break the rules and you don't get a subsidy. Farm size in Europe remains small compared to Australia.

Agriculture's share of Australia's GDP and employment has been declining since World War II. When an economy grows, the percentage of food purchases declines, because it's measured in relation to non-food purchases. We can be sure that the more the world economy grows, the smaller the food contribution will become. I remember a businessman saying as we walked around the Sydney CBD one day, 'There's more money made in these few streets, Patrice, than in the whole of your Australian agricultural sector.' The finance sector generates 10 per cent of our GDP, $140 billion and rising, employing 450,000 people. Agriculture is 2.8 per cent and declining.

For a brief moment in Australia we had a carbon tax on polluters and pollution, which then quickly disappeared due to political hostilities. The echoes of those hostilities can still be heard today in the ongoing energy debate, if a brawl can be called debate.

Then in November 2014 the federal government approved the Direct Action Plan (DAP) to help reduce our GHG emissions. Instead of market-based regulation, emissions reduction would now be via the government's Emissions Reduction Fund.[25] At the time of writing, the DAP has allocated $2.5 billion to pay for 400 million tonnes of carbon abatement in order to meet Australia's 2020 Paris Protocol commitment – the commitment that so many Coalition MPs want to abandon. If all existing projects were deemed equal, and no further funds were allocated, this would mean that the carbon value would be only $6 per tonne, insufficient to drive innovation.

This whole concept is the opposite of a carbon tax, which requires polluters to pay for their pollution. The DAP on the other hand pays the polluters to reduce their emissions. The price a polluter is paid varies with each project, as the government will only fund those where the carbon abatement price is cheapest. The biggest polluters may elect to not participate in the scheme, and some projects are only likely to be developed if the carbon abatement value is high. With a short-term low price, a Renewable Energy Target becomes more important to drive GHG emission reduction in the long-term.

There are funding obligations for the existing DAP projects regardless of new governments and policy changes. The independent expert committee, Domestic Offsets Integrity Committee, reviews and approves all methodologies. Presently there are 22 different methodologies approved in the three categories for emissions:

- agriculture (livestock, soil carbon, fertilisers, feral animals)
- vegetation (regrowth, reforestation, avoided clearing and avoided harvest)
- landfill and alternative waste treatment (AWT).

The register of offset projects, managed by the Clean Energy Regulator has no biochar projects. Most are landfill legacy avoidance projects. Each Australian carbon credit unit (ACCU) is equal to one tonne of carbon dioxide equivalent (CO_2-e). As projects continue, the ACCUs are accumulated and published in quarterly reports by the Clean Energy Regulator, as part of the *Carbon Credits (Carbon Farming Initiative) Act 2011*.

Farmers could design a scheme to offset emissions and actually get money for not clearing. Huge tracts in western New

South Wales have received money via the Emissions Reduction Fund. It will be interesting to learn how much carbon abatement can be achieved during prolonged droughts.

While Australia has a few federal laws to protect biodiversity, state governments try to address this primarily by regulating land and water management. Each state has some form of biodiversity offset to help reduce our losses.

An offset is a rob-Peter-to-pay-Paul idea writ large, accompanied by the hope that there'll be some semblance of balance after the land-clearing has been carried out, the mining done, the urban-growth project completed, or the big event in the name of Nation Building is under way. But no accounting books can balance the losses caused by natural events like storms and fires. On a farm, it's easy to see how one bad agricultural practice can have a swift cascading effect, sometimes within minutes – as when ploughing followed by a storm leads to soil loss.

The perversity of some offsets and Emissions Reduction Fund payments is that what is often being offset was not going to be cleared anyway. If a bushfire rushes through, should the money be repaid?

Queensland is the most recent recalcitrant state, releasing 19 million tonnes of GHG during 2015 courtesy of land clearance. That means 20 per cent of Queensland's GHGs were from the land-use sector. We need to take our eyes off the electricity sector as the only bogey in the climate debate.

It's hard to see how biodiversity losses could have been avoided with the global population growing from 2 billion to 7 billion in a hundred years. (Australia's population, less than 4 million at Federation, is now more than 25 million in 2019.) Something had to give. But making a living from farming doesn't necessarily equate to biodiversity loss. If you haven't put *Atlas of Living Australia* into your computer favourites bar, do so now. This site is a good way to quickly see how diverse our ecology is, why it's a good idea to preserve it, and how you can become a citizen scientist even if you failed biology in high school. What we have is a jigsaw. We recognise pieces, get a feel for where they should go, and when we finish a puzzle and see all the pieces in place we gain new insight.

Every day in Australia, a development application is put before a government body that will result in something being destroyed in the name of progress. It's a continuation of the old attitude of 'If it moves, shoot it. If it's still, cut it down.' The pioneer gene lives on in many rural people today and is still seen as somehow heroic. To clear the land is to tidy it, to improve it, to be the master.

Right now we're trying to understand why we pay so much for electricity – and admitting that power generation is by far our greatest emitter of GHG is the first hurdle.

But before the belated arrival of sanity, we'll be putting up with more 'we need a new coal-fired power station' nonsense. Coal-fired power stations are extremely expensive to build new and to restart, and those that operate now run well below maximum output most of the time. We need new, flexible peaking capability – from wind- and solar-powered storage, pumped hydro and gas – not coal.

With the Hunter Valley choking on emissions from coal-fired plants (circa 1970s), this issue is a hot topic up here. These stations run at just 30 per cent efficiency. But because electricity is the key to modern life and therefore dominates politics, a new coal-fired power station with 40 per cent efficiency is considered by some a triumph.

Most agricultural land in Australia is designated as grazing land. When talking about the capacity of land to rear animals, we use the term Dry Sheep Equivalent, a sort of Greenwich Mean Time for livestock. Animals aren't equal in their nutritional needs. A cow with a calf needs to eat more than does a nine-month-old heifer; goats eat three times as much when lactating, so there's an emphatic difference in the carrying capacity of a farm when it comes to breeding stock and fattening stock.

When it comes to DSE, horses are on top. More than a million horses live in Australia, although only some 221,000 of these are registered, and 400,000 are thought to be feral. Do the maths: a million-plus horses with a DSE of 10, meaning they need the same amount of food as 10 million sheep. And if the horse is in good condition and working, their DSE rises to 18.

In the Hunter Valley, racehorses are huge money-spinners, with prize stallions valued in the millions of dollars. No wonder they get the very best land, and the best water. But plenty of horse types other than thoroughbreds are worth big money. Polo ponies, horses for various show events, campdrafts and pony clubs, even a good old horse for the average rider – can be expensive.

You may be surprised to know that horsemeat is a multi-million-dollar industry in Australia, and that two abattoirs slaughter horses for human consumption and export their meat to Europe and Japan. We don't eat horses here, at least not know-ingly, so most horse protein that stays in this country ends up as pet food, supplied by the thirty-three registered knackeries in Australia. Many of the animals that end up at a knackery are failed or retired professional sport horses, and their manes and tails are used in brushes, including for human heads and in calli-graphy brushes. Horse fat is rendered as tallow and their hides are sent to tanneries. The going price for a carcass is $300.

Horses live in entirely separate kingdoms. They've moved from the transport sector, where we rode them and sat in the carriages they drew, into the gambling sector, where we bet on them, and now we need to re-evaluate where they sit in the land-use debate. All horses need to be included in primary production statistics, especially as they get first choice of the country's first-class land and water.

Like humans, and unlike cattle, horses have only one stomach. People talk of a paddock being 'horse sick', meaning that horses have eaten the same plants for too long and destroyed the paddock's biodiversity. They are very fussy eaters.

In terms of animal status, I'd list the cultural pecking order this way. In descending order – horses, cows, sheep, alpacas, goats. Does look a bit like size matters.

Insects, on the other hand, have a non-existent DSE score. The future of protein in the human diet is certain to include more insects. As a beekeeper, I'm already impressed by insects, but when a fly-farmer can raise $135 million to develop a fly larvae business in order to convert food waste into protein to feed fish, poultry and pigs, you realise that food production today is rapidly changing.

Animal protein is considered to be a higher order of protein than grain protein, because of its energy density. Most grain protein grown in eastern Australia is actually fed to animals in feedlots. The true stars of plant-based agriculture are the fruit and vegetable growers, growing the plant-based food for healthy living, but sadly, fresh fruit and vegetable consumption has been declining.[26] Alongside this is the inexorable rise in obesity. It's another reason that the agriculture and the health sector should share a ministerial portfolio, with success being judged by the number of people kept *out* of hospital due to food-related illnesses.

Most of the land regeneration work that has been carried out to date has been done with government grants or tax credits. Either way, it's been the public who have paid the repair bill. The sale of Telstra helped, and the Direct Action Plan (DAP) is paying now. Everywhere you turn, across the most productive farmlands in the world, government money is being used to repair ecological damage. In the USA, though, there's been resistance. As the Farm Bill was being discussed in Congress in 2018, conservation programs such as the Agricultural Conservation Easement Program were asking to be fully funded to the tune of $500 million a year, otherwise their land restoration work wouldn't be done. US Big Ag squabbled over the half-a-billion dollars.

Regeneration may be expensive but whatever the price, we can't afford not to pay. And the longer we leave it, the higher the bill will be.

Over the past decade, so-called education farms have popped up in Australia, especially in semi-urban areas and on the outskirts of rural towns, with the aim of showing city people and tourists the old ways of growing things and to give kids the opportunity for hands-on contact with animals. These farms are more tourism than agriculture, more social building than farm building.

Permaculturalists have been leaders in this movement, which also aims to show visitors how to lessen their dependence on the supermarket. Thousands of people in Australia have been taught the rudiments of growing and storing food; the various drying, fermenting and cooking techniques that can be applied to many things we grow. The movement is also at the forefront of promoting small-farm produce on social media and using open-source networks for agricultural education. While hardly a threat to Coles and Woolies, they are helping to produce a better-informed community.

But when a working farm's activities include education, with events and visits, new risk assessments are required. Historically all harvest festivals – the first olive or grape crush, the making of passata, the butchering of pigs in autumn, the storing of corn for winter – have been times of celebration in various cultures. When we're asked to do a public event of one sort or another at Elmswood, we too become an education farm. Enter the modern concerns and costs of Work Health and Safety. Visitors can no longer simply wander around at will. There have to be clearly marked routes, with signs and biosecurity banners: KEEP OUT and DANGER and EMERGENCY ASSEMBLY POINT. A person can walk onto your farm, tread on a weed, embed the seed in their boot and innocently spread the problem to the next farm. We farmers obviously do it all the time, but now we talk about the risk. I was reminded of this during shearing, when friends asked to visit. They arrived at the back of the shed and inadvertently brushed against stinging nettles. Our shed might be a classic, old-style and picture-perfect, and it's a long time since the shears went click, mechanical shearing having begun in the early twentieth century, but there are dangers – loose floorboards, a wobbly gate, dim light, sheep running and jumping, shearers cursing, dogs barking, wool to trip over, prickles to pierce.

In the 1930s, Elmswood focused on merinos, but we stopped raising sheep for the commercial wool clip a long time ago. Prices dropped dramatically, making sheep no longer worth the effort. Now our two small mobs of Suffolks and Shropshires keep the tradition of shearing alive. They live close to home and at night are locked up for safety. Well trained, they're easy to muster by bike or on foot.

There are sheep that don't need shearing – the natural shredders like Dorpers – but they're so destructive, dooming olive trees by eating their bark. When we had merinos we needed a dozen shearers; now one shearer does it all. Our wool is tough stuff, and full of grass seeds and a variety of burrs, yet both the Suffolks' and the Shropshires' wool helps prevent erosion. We place it against rocks in gullies to slow the flow of the water. With the drought, nothing has come to wash it downstream, so two seasons' wool is still in place, looking like remnant piles of melting snow. After a deluge it can spread for as much as a kilometre, before slowly disappearing into regrowing grass. It was overgrazing by sheep that created much of the erosion we see today across Australia. Now our sheep are helping in their small way to redress this.

CHAPTER 8

ADDICTION

THE USE OF HARMFUL CHEMICALS IN AUSTRALIAN AGRICULTURE isn't going down, it's going up. Herbicides, fungicides, insecticides and pest baits are used in feed and seeds as well as on soil and plants. Users are required to have chemical certificates to buy them and use them, to have secure places in which to store them, and to have specialised machinery with which to spread them. Add it all up and the chemical industry is a juggernaut employing thousands. Include the money spent on auditing chemical use and compliance, and the sector gets even bigger.

In Australia, chemical companies are clustered with biotechnology companies under the banner of their peak body, CropLife Australia. To them, agricultural chemicals are all about food security and biosecurity. They rarely use the word chemical except to argue that we're all part chemistry, are made of chemicals.

I struggle to understand why most farmers are so enthralled with chemical warfare. As a group, farmers can be mavericks, but when it comes to chemicals they're a mob of sheep; it's not as if only one motive drives all farmers to use harmful chemicals. Their use is promoted everywhere a farmer turns. Flyers fall out of our letterboxes. Promo cards are propped up at cash registers in supply stores. Chemical 'specials' are always on. They're advertised on the four-wheel drives of stock and station agents parked in town. Chemical companies sponsor everything. You'd have to be deaf, dumb and blind not to notice it.

In *Behemoth: A History of the Factory and The Making of the Modern World,* Joshua B. Freeman reminds us of the unstoppable rise of entrepreneurs wanting to better exploit their workers. The first factories were born out of people value-adding to primary production. Cotton millers, sausage makers, flour millers, bakers. By 1929 the Ford factory at River Rouge in the USA employed 102,811 workers. In the modern era we see the results of this trajectory at China's Foxconn Factory at Shenzhen, with a work force of hundreds of thousands, and some buildings wrapped in yellow netting to discourage suicidal jumping.

The same upscaling has been applied to farm animals. In Australia today we have beef feedlots with a capacity for 65,000 head of cattle. The animals may have been born in a grassy paddock but more and more spend the latter part of their lives in a feedlot. Sometimes they spend a hundred days there, but many specialist beef producers imprison cattle for three hundred days, almost a year. All feedlots are confined spaces, with a mixture of yard shapes and sizes, and long troughs where the feed is served. A mixture of high-protein grains, roughage (such as nut shells or cottonseed meal, because the cattle's diet is around 60 per cent fibre), vitamins and antibiotics. Perhaps this was predictable. We shove our elderly into nursing homes at the end of their lives – why wouldn't we do it to animals?

The feedlot industry has seen rapid global growth everywhere meat production is carried out. Not just for cattle, but for sheep, pigs, chickens, ducks, turkeys and fish. In Australia in 2017, 2.9 million cattle were sent to a feedlot before slaughter, a 20 per cent increase from 2007. The beef feedlot industry (now a formal subgroup of the wider beef sector) claims that in 2017 they contributed $4.4 billion to our GDP. Edible beef by-products, such as tallow, bonemeal, gelbone and beef extract aren't identified by provenance, but may well come from lot-fed animals.

Feedlots have grown because consumers demand cheaper meat all year round, and this industrialisation of production drives the food industry's chemical addiction. Modern food systems don't work without them. Feedlot cattle are made to stand all day in shit, covered in flies, while being force-fed specialist feed mixes with antibiotics to keep the bugs under control to make them fatter faster.[27]

While I'm an organic farmer, my cattle can still end up in feedlots if that's the next owner's decision. Ethical farming means little when organic and non-organic supply chains merge. Fortunately, more and more people are speaking out about the abuse of farm animals, this violence against nature.

And it is violent, and so can working in such a place be. Years ago, Scone had a piggery, where sows were tethered and living appalling lives. A worker was doing his daily rounds, feeding them grain with antibiotics, when a sow lashed out and bit him on his scrotum. The pain was horrific, his scrotum swelled. The young man, not long out of school, was rushed to hospital, where an infection developed. His testicles had to be removed. There's a tragic irony in this story of a mother pig attacking the young man's manhood. The ultimate pig statement about life in an industrial piggery.

With so many farm chemicals on the market, the research on their negative effects on humans can't keep up, despite the many compliance rules in place. Tests don't cover the full exposure or the chemical mix an individual may receive, nor do they isolate precisely where a particular chemical has come from. The government conducts random tests for contamination, but that's precisely what they are – random. And 99 per cent come up clean by Australian standards. Should we feel good about that result? Not when the tests are being conducted for something known not to be there. I'd be happier with a press release from CropLife proudly announcing they're selling fewer chemicals.

When hay is delivered to drought-stricken properties, some farmers also buy alternative feed to keep their stock alive, such as grape marc and waste from the confectionery industry. Who knows if these unusual dietary additions affect the flavour of the meat or its nutrition. While perhaps okay, they've not been tested as animal feed.

Organic farmers know about building their soil with the power of fungi. In biodynamic farming this is done with the mixture we call 500: cow manure that's been buried over winter, then diluted with warm rainwater and mechanically stirred, before being sprayed on soil to feed the microorganisms that look after the soil in return.

Many conventional farmers also understand that microorganisms make good soil structure. There are bad bacteria and fungi as well, of course, which can be a problem during major surgery, for example, and can even cause pandemics. But we're yet to fully understand their complex role in soil health.

This is something biodynamic farmers have always talked about, not using scientific names for microorganisms, but using observation. Based on my farming observations, I accept known unknowns such as microbes playing a mysterious but crucial role in making soil function. They build soil by breaking down biomass, such as leaves and twigs and manure: trillions of complex molecules marching across all soil types all the time. They cycle nutrients releasing them for plants to use. Studies reveal that soils have off-the-charts complexity across different biomes. Just as coal was once vegetation, soil was once plant matter, and during every second of every day microbes control the transformation, both using and storing carbon in the process.

Getting a basic soil test done before mechanically spreading elements on it doesn't always work. Soil has to be biologically active with fungi and bacteria if it's to make use of any extra elements we feed it. It's like putting make-up on someone who's really ill: it might disguise the pallor but it doesn't make the patient well. Just as the human gut needs microbes to maintain health, so does the soil.

One year, I hesitated to plant garlic because of drought. Beneath the wisps of dry grass, the parched earth looked incapable of supporting anything. Graeme didn't agree, remembering how fine crops can grow after prolonged dry periods because, he believed, the myriad insects that die in the soil provide a nitrogen boost. His theory may have an element of truth, because research now suggests that dead microbial bodies may be the most persistent form of soil carbon. When considering all the carbon released by agriculture, it's comforting to think about these invisible microbes.

The use of antibiotics in agriculture has particular implications for human health, since it increases our resistance to them, thereby making them less effective. While this is an ongoing discussion, full of claim and counterclaim, the fact is that many feedlots use antibiotics in their feed mixes. The producers insist that their animals manage very well with the low antibiotic use, but it's people who are having the problems with infections.

There's danger in every cowpat. Animals shit all over the place on a farm, and their faeces are eventually converted back into soil, thanks to microbes and dung beetles. The grass grows again and animals eat it. In every little pat plopped on the soil are the residues of the drugs administered to the animal along the way, leading to profound changes in the soil, and in humans.

Users of synthetic chemicals wear protective clothes while spraying, or they sit in air-conditioned tractors believing they are somehow separate from the nature they are poisoning. They like to think that all the chemicals they apply are somehow absorbed into the atmosphere and soil and water with no effect. Why then are they so secretive about what they do? People who use chemicals never boast about it. They don't Facebook it. Or take selfies in their protective gear, proudly showing off their poison. You don't see them spraying chemicals in TV food ads. Theirs are not the farms having open days, yet they are the farms most ministers for agriculture believe are the foundation of our agricultural sector. Real farmers. Modern farmers. It's the small farmer who's likely to have open days, to educate people about animals and rural life. Sadly, most domesticated animals are not living their life in nature on small farms.

In 1989, the National Industrial Chemical Notification and Assessment Scheme (NICNAS) was set up in Australia, to provide the nation with 'safe and sustainable use of industrial chemicals', not just those used in agriculture, but everywhere. NICNAS claims there are 38,000 approved for use. In 2005, when the EU started testing the effectiveness of thousands of farm chemicals, they found that many didn't work, some were unsafe, and most had never been tested for environmental safety. Soon after NICNAS announced that only 125 'existing chemicals' had been assessed.

The charter of NICNAS includes upholding the principles of the community's 'right to know', as identified in the

Bahia Declaration of the 2000 Intergovernmental Forum on Chemical Safety. So you'd think it would be easy to find out the volume of a chemical sold. But it's not. When the EU's European Commission decided that neonicotinoids were bad for bees and voted to restrict the use of three neonicotinoid compounds (imidacloprid, clothianidin and thiamethoxam), Australian farmers and beekeepers made their case. Despite there being no countrywide survey of the status of Australian bees (beekeepers are not even registered in the Northern Territory), the government declared that Australian bee populations were not in decline, and the Australian Pesticides and Veterinary Medicines Authority declared the chemicals safe, even though they hadn't conducted an Australian review into neonicotinoids.

If we are to have agriculture, we must have bees: food needs pollination. Bees are complex livestock and often dismissed because of their tiny size. The Australian Honey Bee Industry Council, the peak body for beekeeping, assesses the value of honey and bee products sold at $90 million annually. Yet we can't manage to assess a chemical that the EU believes is killing them.

Every time we sell or move cattle at Elmswood we must sign a document, a National Vendor Declaration, stating whether the animal has been treated with a synthetic chemical or drug or has eaten feed that has been so treated. Is the truck clean? Tick tick tick. We sign away. This means that every animal entering the market, to eventually find its place on people's plates, has been declared at the point of sale to have or not have been in direct contact with chemicals during the last few months of its life. Here are some of the other questions on the declaration:

Have any of the cattle in the consignment ever in their lives been treated with a hormonal growth promotant [HGP]?

Have the cattle in the consignment ever in their lives been fed feed containing animal fats?

Has the owner stated above owned these cattle since birth?

In the past 60 days have any of these cattle been fed by-product stock feeds?

In the past 6 months have any of these animals been on a property listed on the ERP [Enterprise Resource Planning] database or placed under restrictions because of chemical residues?

Are any of the cattle in this consignment still within a Withholding Period [WHP] or Export Slaughter Interval [ESI] as set by APVMA or SAFEMEAT, following treatment with any veterinary drug or chemical?

In the past 60 days, have any of the cattle in this consignment consumed any material that was still within a withholding period when harvested, collected or first grazed?

In the past 42 days, were any of these cattle: A) grazed in a spray risk area? B) fed fodders cut from a spray drift risk area?

On the back of the National Vendor Declaration booklet is a list of all the chemicals that can be legally used, and their withholding days, the amount of time before an animal can be legally sold. In the case of Young's Triclamec Cattle Pour-On, and Fasimec Pour-On, both require cattle for domestic consumption to be withheld for 49 days; if they're heading overseas, a with-holding period of 140 days is required. Australian and overseas regulations show variations, but it's the overseas standards that are higher.

There are similar questionnaires for fodder. Has a hay crop been grown on a property with either an organochlorine status classification or under quarantine because of residues within the past twelve months? Has the fodder crop been subject to spray drift during its production? The National Livestock Identification System and the National Vendor Declaration show how much chemical use has been normalised throughout animal food pro-duction. The fact that all farmers, including organic producers, must fill out these forms reveals the extent of chemical use and the risks and concerns the government has had since the 1960s, when these forms were introduced.

Australia does have the *National Residue Survey Administration Act* to help keep our industrially manufactured chemical usage under control, but there is an alternative, and that is not to use them in the first place. Even a former governor-general, Michael Jeffery, has dared to suggest, as president of Soils for Life, that we should use fewer synthetic chemicals and inorganic fertilisers.

Weeds are another target for chemical users. They're like body odour for some farmers, who judge each other on how many weeds they have or haven't. They allocate marks for varieties. Blackberry is big bad, verbena less bad. White settlers introduced many plants we now deem weeds. We're still doing it. I'm a devoted subscriber to garden clubs and online plant sales, but another reader will say in horror, 'Did you see that they are still selling *Verbena bonariensis*? How are they allowed to do that?' Verbena is a weed to some of us. Poisoned here, planted there. We have given up capital punishment, wearing gloves to town, at-fault divorce, and we've survived. The same acceptance of change must apply to plants. We need them all in this climate emergency, even weeds. All plants have a place and purpose. While some are annoying, they don't deserve chemical warfare, although if we know a plant is likely to become a garden escapee it's our duty not to plant it, especially in a rural area.

Farmers are growing the food plants we want now, but the chances are that future generations will think we were lunatics. Who knows, maybe quinoa will one day be more popular in Australia than wheat? And people will be poisoning wheat because it's a weed. I have a friend who loathes wild oats and barley grass.

While hazard reduction burns and firefighting are important jobs of the NPWS, they're also fully trained in the dousing of poisons to control weeds. But there are biological controls for some of our nastiest weed infestations, such as tiger pear, *Opuntia aurantiaca*. This small cactus, native to Argentina and related to the prickly pear, is rarely higher than my knee. It was apparently introduced to Australia as a pot plant, which is a bit like putting a piranha in your goldfish bowl. They do have beautiful yellow flowers, which we can spot from afar when they flower majestically but delicately around Christmas time.

Despite their beauty (I've never known a plant to trigger such emotion: horror and wonder), Phillip and I will get up before dawn to hunt them down. When they're not flowering they can be difficult to see, hiding in the grass and in the nooks and crannies of fallen timber. A drought will reveal more tiger pear than usual, and kangaroos unwittingly spread them everywhere, the spines sticking in their fur until the seedstock bounces loose; we're forever finding new clusters of tiger pear along the fence lines. To spread the biological control that kills them we carefully break off a piece of tiger pear infected with the cochineal insect, *Dactylopius* sp., when it's most active and carefully transplant it into a healthy plant in a different area. A cochineal infestation looks like white marshmallow spreading across the prickles. Warm temperatures and dampness can trigger the rapid spread of cochineal, but balmy winters and chilly mornings don't stop the cochineal coating the tigers. We don't actually collect the tiny cochineal scale insects.

Different species of cochineal are host-specific. And the different *Opuntia* species surrender to different cochineals. The biggest cactus we have is rope pear, *Cylindropuntia imbricata*, which has a mass of magenta flowers in summer. They can be the size of a four-wheel drive and become impenetrable. Its particular cochineal control, *Dactylopious tomentosus*, will spread right across the plant, almost killing it but not quite. That's the old thing about hosts and parasites: they need each other and so coevolve, and the cacti adapt to attacks and near-death experiences. This cochineal is controlling the rope pear, but not eliminating it. Whereas tiger pear, once attacked, withers into oblivion.

The moth *Cactoblastis cactorum* has been keeping the famous prickly pear, *Opuntia stricta*, under control for decades though something has changed during this 2017–19 drought. Neighbours agree the pear is regaining ground. An army once thought defeated is on the march again. Under our olive trees thick, juicy pear plants are popping up everywhere and we're on high alert to see if biological control will be enough again. The moth, when in the caterpillar stage, munches inside the cactus pad and before we realise it the plant collapses.

It's a paradox that the loathed prickly pear is considered a plant of great importance in the Americas. It's on the Mexican coat of arms and is used as a natural fence line; it forms a cactus curtain around the infamous Guantanamo Bay in Cuba. While we might hail the benefits of *Cactoblastis*, this biological control for us is unwelcomed elsewhere.

Biological controls can also go very wrong, of course, as the invasion of cane toads dramatises, but this shouldn't stop us deploying biology as a measure against aggressive plants, insects and animals. Poisoning with synthetic chemicals is worse.

Many chemicals once deemed safe have been reclassified as high risk and deregistered. Others have been found to be less safe than first thought, yet are still widely in use. Perhaps the best-known example is glyphosate, via its use as the basis for Roundup, which is readily available in supermarkets and hardware stores.[28]

Glyphosate might sound like an energy drink but it's the world's most contentious chemical. In 2018 a court in Los Angeles found that a school employee had developed cancer because of handling the chemical, and a jury awarded him $289 million. The manufacturer of Roundup, Monsanto, demanded the court throw out the verdict. It didn't, although it did reduce the punitive damages. Now nine thousand people have signed up for similar lawsuits across the USA.

At the time of writing the Australian Pesticides and Veterinary Medicines Authority continues to approve products containing glyphosate that's present in hundreds of different products registered for use – including Roundup; though in 2015 the World Health Organization's International Agency for Research on Cancer warned that it's 'probably carcinogenic in humans'.

The maximum level of glyphosate allowed in grains entering Europe has been reduced, and the EU is moving towards a total ban. Glyphosate is already banned for use in public spaces in the Netherlands, and France banned Roundup Pro 360 for use in January 2019, effective immediately. President Macron has said all the other glyphosate products will be outlawed in France by 2021.

This is challenging for many agricultural producers in Australia, especially grain growers and vignerons. A headline in *The Land* in 2018 read: GLYPHO BAN 'DISASTROUS'.[29]

The accompanying article went on to quote local farmers as saying a ban on the chemical could 'send our agriculture back to the 1980s'. Some scientists argue that a ban will cause more soil erosion, because farmers would have to revert to heavier ploughing, or be forced to use even worse broad-spectrum chemicals. At one meeting, I heard farmers blame the worker who'd sued Monsanto: 'it was his fault for letting himself get drenched in the chemical in the first place'. When the Munich Environmental Institute found traces of glyphosate in fourteen different German beers in 2016, no one blamed the drinkers.

We already know how to farm without this chemical, as evidenced by all the organic farmers successfully growing grain,

meat and cotton. For instance, steam can be used to kill weeds, as can paddock rotation systems designed to build organic matter. I've sometimes used a flame gun to burn weeds, although with fire bans being extended, this option is thwarted. So we're putting more effort into trying to manage the pasture and garlic beds with strategic grazing. Sheep love wireweed, as we discovered after garlic harvest one year. And cattle will eat saffron thistle when it's small. The concept of no-till organic farming might sound like an oxymoron, as no-till farming is chemical farming (paddocks are sprayed with herbicides to kill weeds instead of ploughing before sowing a crop) but work in the USA has proved you can get healthy, high-volume green manure crops by using machines specially built to cut the crop when mature and plant seed at the same time, in a process that leaves a heavy carpet of mulch over the land. Pasture cropping, a system designed by Col Seis and Darryl Cluff in New South Wales in the 1990s, is similar whereby winter crops have been successfully sown straight into native perennial grasses. This way there's no bare earth and less opportunity for weeds to thrive.

Transitioning away from glyphosate will take years. Monsanto (now fully owned by Bayer) will fight every case brought against it. When the ABC's Josie Taylor reported in 2018 that local councils were using glyphosate widely, she was accused by the chemical industry of sensationalism and fearmongering.[30] But councils across the country, including the Upper Hunter's, are at least debating how to manage without glyphosate. I'm eager for the day when we'll look back at this period with incredulity and disbelief – this period when families can put Roundup in their supermarket trolleys along with their groceries.

The lesson here: don't believe a chemical company when they say their products have been tested and proven safe. They would say that, wouldn't they?

But even with stronger regulations in place, things can still go wrong. In April 2018 it was discovered that an Australian barley crop sent to Japan had residues of the pesticide azoxystrobin at five times the allowable limit. Some of the barley had already been distributed, and used in food products or consumed directly, before the discovery was made. A voluntary recall of the remaining products was made, and the episode triggered the Japanese government to watch grain imports from Australia far more closely.

Blueberries is another crop that get doused in chemicals. National production of this fruit has more than doubled over the past five years; we now grow over 11,000 tonnes per year. And during the past thirty years, the price of Coffs Harbour real estate has been boosted by this fruit. A 40-hectare block suitable for blueberries can now sell for over $1 million. The blueberry boom is now far bigger than tourism at Coffs and contributes a quarter of the $1.5 billion that flows into the local economy from mixed farming each year.

Not all the locals like the blueberry invasion. There's deep concern about increased chemical pollution. A lack of regulation has allowed the new arrivals, who tend to be the heavy users of sprays, to set up next to schools. There are also complaints about the ugliness of the netting, the pollution of waterways, and the large dams being built without approval. In the Hunter the signs say NO NEW MINES. Around Coffs, trees carry signs saying NO BLUEBERRY FARMS.

I know what it's like to see dramatic, almost overnight changes to your local environment. You're confronted, jolted, don't know where to turn. Although the Environment Protection Authority has the right to monitor blueberry operations, cuts to budgets and staff limit their capacity to do much. Australians demand control of our boundaries, some demanding border protection from desperate refugees – people can feel the same way about miners and industrial food producers.

While regional centres in Australia try to adjust to industrial agriculture, some old cities, like Paris, are asking farmers to come back closer to home and is encouraging the development of 'urban agriculture', to provide more fresh food in the city via pop-up farms. Think portable raised beds on industrial wastelands. These won't of course replace supermarkets, but they do provide a healthy choice for shoppers. Many involve new ways of waste management, such as using restaurant detritus for compost, coffee residues for worm farms. Such sustainable planning could help produce harmony in Coffs, to the advantage of mixed farmers, including blueberry producers, the locals and the tourists. Everyone can win.

Meanwhile it's a standard claim of companies that using their fertilisers will make plants stronger, so that they need less pesticide. And they'll urge farmers to use their particular sprayer because coverage will be better: 'reduce your chemical volume by 99 per cent!' There's an odd logic at work here: these companies, after selling chemicals they claim are essential, then trumpet a new product on the grounds that it reduces chemical use.

That said, there are new-generation sprayers that use artificial intelligence to recognise specific weeds, insects and diseases in a crop. They do this from a height of half a metre and they then target the pests with a droplet suspension system, a technology born in the medical sector. There are already a few targeted droplet dispensing systems on the Australian market, which work with varying success. But the price of such machines will mean they're only for the corporate farmer, as are the mega-harvesters and planters. The price of an overhead olive harvester is already $500,000, way out of reach for a small-to-medium grower.

Chemical companies make dubious claims of reducing poverty by selling us better seeds and more fertiliser. They want us to forget the pre-synthetic chemical agricultural heritage built over millennia – traditions and methods specific to a thousand unique communities and millions of different ecosystems, nurtured by the creativity of farmers, all doing fine without the bombardment of homogenised commercial formulas – and accept instead a chemical corporation-dependent solution to a problem that doesn't really exist.

While there are scientists and statisticians trying to unravel the impact of the multitude of chemicals that humans encounter in their daily lives, it's difficult to isolate and identify specific risks. A population-based cohort study (a type of research that looks at large population groups of approximately similar age over a long period) assessed 68,946 French adults (78 per cent of whom were female) and concluded that those who ate organic food had less cancer.[31] The result will drive the next wave of research, to confirm these findings. Eventually science will isolate further which foods and chemicals are linked to which cancers.

As well as contaminating soil and water, synthetic manufactured chemicals leave behind mountains of plastic containers. They end up anywhere and everywhere. Furtively carted across distant paddocks to be hidden in gullies. Or buried in holes and forgotten. Dirty little secrets to be discovered decades later.

Back in 1999, drumMUSTER began to collect empty chemical containers, and now there are more than eight hundred official sites across the country where containers can be deposited. A chemical user must book in their used containers, and there's a register of the number collected. In 2018 drumMUSTER announced that it had collected a total of 32 million chemical containers since 1999. All drums are voluntarily surrendered. That number equates to 32,000 tonnes of plastic not going into landfill. Plastic containers are recycled to make new plastic products; pipes, fence posts even your wheelie bins.

ChemClear is another group doing a similar task, managing leftover chemical disposal. Both groups are part of AgStewardship Australia, a non-profit organisation funded by levies from chemical manufacturers and supported by farming associations, community groups, Landcare, and state and local governments. Unused toxic chemicals are incinerated at high temperatures, up to 3500 degrees Celsius. This rearranges the chemistry and takes the ingredients back to their original building blocks. Some leftover liquid chemicals re-enter the world via industrial waste-water treatment plants. Some are so toxic that the only thing to do is reseal them and bury them. A lot end up in the cement-manufacturing sector as a source of energy.

I liken the end of industrially produced synthetic chemical use to the end of slavery. The anti-slavery crusaders of the nineteenth century had to face down arguments that abolition would mean the end of the cotton industry and the food bowl of the Deep South.

Today the slaves may well be the modern farmers themselves, and the manufacturers of industrially produced synthetic chemicals their masters, because the only ones profiting through droughts, floods and fluctuating markets have been chemical companies. The government approves their products, our taxes pay for courses on how to use them (Local Land Services trains people how to bait, for instance), compliance and auditing and allows the same companies to sell domestic variations of the chemicals at higher prices in supermarkets and hardware stores. And the public also pays for the clean-up of the poisonous residues.

All this as the word 'sustainability' is slowly being phased out in favour of the term 'cyclical economy'. I hope it's not a tread-mill. Bad ideas need to be thrown off.

In my future world I'd like to see rural-supply stores only selling non-polluting, safe green chemical products that you don't need a certificate to buy or a place in which to lock them up. And in that future world there won't be a need for mandatory nutrient fortification of our food because we'd be able to provide whole-some raw ingredients for our food. Clean and green won't be a marketing tool used by chemical companies, but a reality.

CHAPTER 9

SWEET TRUTH

THE BENEFACTIONS OF BEES ARE MANY, INCLUDING THE ALCHEMY
of turning nectar into honey, the pollination of countless food
plants. On weekends, when the sun's shining on the patio and
Phillip's returned from the Gundy store with the papers, I'll make
a pot of coffee and we'll do the general knowledge quiz in the
newspaper. We do fine with geographic and political questions,
fail utterly at sport. After one of these morning rituals we pack
our old truck to head for the hills, specifically Gloucester. The
two-hour drive to Gloucester is one of our favourites. It's part of
an official tourist loop but most of the road is deeply rutted, and
in winter is covered in snow. Signs warn off drivers in small cars
and anyone towing a caravan.

We'll enjoy the day at the Gloucester market, collecting a few
weeks' supply of our favourite biodynamic yoghurt and cheese

made by friends David and Heidi. Plus loaves of Fosterton bread made from biodynamically grown grains, some jams and vegetables and hopefully some unusual pot plants.

Phillip loads an extra spare tyre just in case. We drive through the pretty little township of Moonan Flat, the road then winds and climbs steeply, first to the Dingo Gate and then the spectacle of the Tops, as the Barrington Tops National Park is affectionately known. We stop for a thermos tea break, enjoy a brief wander through the cycads and ferns, and pick a few mushrooms. Air doesn't get any fresher, vistas more awesome. We're back in the dawn of time.

Climbing through the Tops we come to a wide, sweeping corner and see hundreds of rainbow-coloured beehives. A Manhattan of hives. But something is very wrong and we stop to investigate.

It looks like the aftermath of 9/11. The boxes are tilted askew, some crushed. What happened? Is this storm damage? Refugee bees are trying to make sense of the chaos, clustering outside the wrecked hives. Bees have a genius for repair work but this is way beyond their capacity. Then we see tyre tracks. This is deliberate. This is vandalism. A truck has rammed into an apiarist's city of bees.

I walk around taking photos, and an hour later at the Gloucester markets I make a beeline for the local honey stand and show the keeper my photos. Every beekeeper in Australia has an ID number and is expected to have it on all their hives. But these tipped over hives are hiding the number, so we can't identify the owner, but the keeper at the honey stand and I think we know the owner of the land. I make a few calls and describe the tragedy. What could possess someone to wreak such havoc? To destroy so many hives in spring, when the colonies are building for their summer splurge, is particularly criminal. And millions of bees will be forced to seek refuge in trees in the National Park, and thus become wild bees.

Every season, beekeepers are caught up in a native-bee versus honey-bee debate. The honey bee is not indigenous to Australia. Its presence dates from around the 1820s, and though these bees are officially allowed to live in our public parks, where some of the best honeys in Australia are made – kauri, jarrah, leatherwood – many ecologists argue that they shouldn't be in national parks. Before the arrival of the honey bee, native vegetation was pollinated by native insects, including the many different solitary native bees. The social honey bee, on the other hand, is more aggressive and in its small way is changing our bushland. Bees can leave our managed hives and set up in tree hollows. It's been suggested there are now more wild bees in Australia than managed bees.

We never learn if anyone was found responsible for the crime in the Barrington Tops, nor do we learn the motives of the attack.

Gavin was a beekeeper. He arrived at Elmswood fully qualified to deal with all creatures great and small. A beef and lamb producer, he'd also planted an olive grove. I've met few people with his skill set, before or since. Our young manager John is mechanically sophisticated, has less experience with stock and none with olives, but thankfully studied bees at Scone High School.

I move our hives constantly across the farm, seeking protection from weather, cattle, wildlife and thieves. Bee rustling is a problem here too. I try to find locations convenient for both the bees and me, and where they're not likely to be kidnapped. After planting Silverado lucerne on the recommendation of a local queen breeder, I decided to bring some hives close to the crop, which was right next to the garlic. Nothing warms the heart more on my daily walk to the garlic patch than seeing the bees busily going in and out of the hive. You can learn a lot by their behaviour at the hive entrance. The volume, direction and speed of bee movements indicates level of nectar flow, full pollen sacs on their tiny legs show protein supplies are being delivered. As the boxes fill with honey the aroma can be dizzying.

But the hive's closeness to the garlic led to one of our casuals being stung, and suddenly my bees were Apis non gratis. Workers

were nervous, distracted. Perhaps the culprit wasn't one of my bees, but a fierce feral, but regardless, we had to move the hives.

John's fearlessness with bees and his ability to interpret their behaviour are a big help to me. 'Bees look like they're getting crowded in the tree-lot hives,' he'll advise. Or, 'Lots of pollen being collected.' He has the makings of a fine apiarist.

We have hives at Elmswood because I love bees and we love honey. The last farm job I'll give up will be beekeeping. We'll cut loose the last few cows, uproot the olive grove and the garlic patch, but don't dare threaten my bees. Bees may be small but their collective IQ is so high they could join Mensa. Alchemists of old wanted to turn base metal into gold, but bees' production of liquid gold from flowers seems as ingenious. And every jar of fine honey is different, depending on the seasons and which trees are in bloom.

With food fraud as big an issue as fake news it was only a matter of time before the news of fake honey made headlines. It was discussed frequently within the industry, highlighted at conferences and reported in beekeeping journals. Deception occurs in many products, not just with honey. When the Australian Olive Association set up standards for extra-virgin olive oil, it was soon revealed that many imported brands were not in fact extra-virgin, and some weren't even olive oil, having been blended with cheaper vegetable oils, such as canola. The outcome of those revelations? An almost total lack of interest in the fraud by the federal government and the supermarket chains.

Substituting, adding, tampering with, or misrepresenting

ingredients for economic gain happens in many food categories. Horsemeat can be presented as beef. Peanut shells are added to ground cumin. The synthetic replaces the authentic. With vast volumes of produce and ingredients being transported over great distances, the opportunities to fake multiply. It's all too easy to be conned.

Nor is the food industry the only culprit. Forged spare parts became a big problem for airlines. They even once found their way into Air Force One, the US president's 747. When accidents occurred, the aviation industry was forced to improve its systems and quality controls. Scandals force car manufacturers into expensive recalls. But it's hard to get governments to worry about corrupted olive oil or honey. Inspection systems are costly to administer and there's the loss of revenue to consider. We're often more concerned by tax losses than the fact that countless consumers had bought honey that no honest bee would approve. And here in Australia, it's now been proved in laboratory tests that sugary syrups imported from China have ended up in famous brands on the supermarket shelves.

While these crimes don't stay on the political radar for long, food fraud does matter, whether it's the melamine in milk that caused widespread deaths in China, or the polluted olive oil that killed many in Spain. Or the listing of false ingredients, false statements, false labels, counterfeiting adjustment of expiry dates, the redirection to the commercial market of emergency relief shipments intended for NGOs in areas hit by natural disasters. Or contaminated food sold as organic.

Governments do set minimal standards, but it's impossible to keep watch over all local production, let alone imports and the criminal mind. We try hard with intelligence gathering, laboratory testing, auditing, and sometimes with criminal charges, but just as the system failed in the War on Drugs, so it can with food. As long as there's a profit to be made, food fraud will flourish.

So if it looks like honey and tastes like honey but isn't honey, what is it? Liquid sugar made from corn or rice, most likely. Australian beekeepers also use sugarcane in their bee management; sometimes when nectar is low, winter is harsh, or queen rearing is in full swing. When you feed bees sugar, they'll forget the flowers and obediently convert the sugar and place it into their combs. A beekeeper could then extract a syrup that looks like honey but has nothing to do with local nectar.[32]

Adulteration with sugar threatens the credibility of the whole industry. In some international tests 40–80 per cent of honeys were found not to be honey at all. And with global production slowing, rising prices are an inducement to bypass the beehive. (The fall in production has been worsened by the phenomenon collectively known as colony collapse, where hives have been dying or bees disappearing for complicated reasons: the use of bad chemicals, poor nutrition, infections, viruses, or a mix of all these.) The only safe way is to buy honey from a local apiarist who knows their plants. And it's best to avoid honey that seems very cheap. If it's cheap honey it's probably funny honey.

During its brief life of about six weeks, a single worker bee collects enough nectar to make less than a teaspoon of honey.

Billions of bees are needed if more of us want to enjoy real honey. Apiarists aren't simply honey producers any more; many focus on pollination, some breed queens, some sell whole hives, others design better boxes. Australian honey bees are tough and we're yet to be inflicted with varroa mite, a bug that has almost decimated beekeeping on every other continent, so it's no surprise that our bees are being exported as the global bee count diminishes. Yes, live bee exports. But they travel in luxury.

For those of us who focus on honey, pure and simple, there'll be a growing differentiation between honeys. The highest anti-bacterial honeys will soon be kept in the medicine chest. There'll be honey for toast, honey for a sore throat and honey for serious infections. That's of course so long as we don't kill off our medicinal nectars. When driving to the farm along the M1, I have a little sob at the clearing of the bush along the median strip separating north- and south-bound traffic. All the leptospermums and the nectar-rich shrubs are being bulldozed and chipped to widen the freeway. Development? Progress? Or vandalism? Proudly proclaimed as native regeneration just thirty years ago, these same trees are now gone. There'll be consequences.

In 2016 I spent a magical day at the Newcastle botanic gardens, where the Hunter Amateur Beekeepers Association keep their hives and meet each month. The day was training for future tastings and honey-judging. While I've entered Elmswood's olive oil in agricultural shows and proudly won a ribbon or two,

I've never ventured into a honey competition. With every jar unique, how could I decide what to enter? And there are so many honey criteria:

Class 1: Liquid Honey 'International'; Class 2: Liquid Honey Manuka MGO Range 263–999; Class 3: Liquid Honey Manuka MGO Range 83–262; Class 4: Liquid Honey, Standard Colour WHITE; Class 5: Liquid Honey, Standard Colour LIGHT AMBER; Class 6: Liquid Honey, Standard Colour GOLDEN; Class 7: Liquid Honey, Standard Colour DARK; Class 8: Creamed Honey; Class 9: Granulated Honey Fine Grain; Class 10: Granulated Honey Coarse Grain; Class 11: Comb in Liquid Honey (Chunk Honey); Class 12: Pollen, cleaned and dried; Class 13: Cut comb (Comb Honey); Class 14: Beeswax – White; Class 15: Beeswax – Yellow; Class 16: Beeswax – Tablets White or Yellow; Class 17: Mead Sweet; Class 18: Mead Dry.

There is also an official colour chart, with shades from light through golden to dark. Pure honey varietals have a consistent colour range, but for people like me who keep bees close to home and extract honey that is naturally blended by whatever is in bloom, there's always a wide variety in viscosity and seasonal texture.

For our taste-test lesson we turn all our sample jars upside down, then turn them back upright and note which honeys turn the quickest. The thicker the honey, the slower the turn, and the higher the point score. By day's end we've sampled and turned so many I feel like a bee myself, my tongue having travelled all across New South Wales licking flowers. Some honeys are like sipping sweet blossom, with the aroma as strong as the taste; some have a bite at the back of the throat; others are woody, taking you into a forest. My favourite Elmswood honey is always the first taste at a fresh extraction, when the wax coating is broken and you chew a little waxy honey and feel the sweetness explode in the mouth.

Weeks later, the group meets again to discuss mead – alcohol created with honey. We mix various proportions of honey, water and old-fashioned herbs according to ancient Latin texts. Mead is an obsession of the teacher of our group; she's been experimenting with old recipes for years, and it's easy to share her excitement. Sipping her meads is like sipping wine; we taste the sweet and the dry as she describes how to age them.

Honey-tasting hit new heights for me when a colleague from a truffle farm near Braidwood sent a gift: a coastal honey blended with grated truffles. It was like nothing I'd ever tasted. Excited, I took the jar to the garlic team so we could all have a sample. Everyone gave it a gold medal.

To test everything on the farm would cost a fortune, so we're yet to test our honey for its bioactive qualities. The famous manuka honey flowers, *Leptospermum scoparium*, grow profusely along Australia's east coast. Plenty of other *Leptospermum* species have been tested for their high bioactivity effects too, and some are much more impressive than *L. scoparium*. And it's not only *Leptospermum* flowers that have these medicinal qualities. Those of the magnificent jarrahs of Western Australia, *Eucalyptus marginata*, also have antibacterial properties. Eventually I'll test our honey for its methylglyoxal (MGO) level, the quality that defines a honey's medicinal power, but it won't add to the bottom line. We don't collect very much to sell and we leave most of it for the bees; it's their food too.

In 2017 I undertook the New South Wales DPI's Certificate III Beekeeping Course at Tocal College. Most of the group were already beekeepers. The training is part of NSW DPI's commitment to professionalising the industry with a focus on biosecurity.

Photo courtesy of Steven Honeywood

CHAPTER 10

REARRANGEMENTS

RURAL AND BUSINESS MEDIA BANG ON ABOUT STAKEHOLDERS (primary producers) being the backbone of the nation, their rural communities poised and prepared for inevitable and necessary change. A billion dollars-plus in drought relief for New South Wales in 2018 would suggest otherwise. What I see is fear.

We've had more than our share of political heavy-hitters elected up here, two former agriculture ministers and deputy prime ministers. John Anderson helped usher in our first major water reforms in the 1990s, only to move on to become a land developer. More interested in energy than agriculture, he put out the welcome mat for coal-seam gas and recently the inland railroad. In 2019, Barnaby Joyce is still spruiking the desperate need for more coal-fired power stations. With 304,200 people employed in agriculture and 85,681 primary production businesses across the country, government agricultural institutions, especially local ones are getting less and less support. Even though drought money can be found. In 2016–17, the LLS was allocated $159 million: $4.2 million less than the year before. These cuts mean policy doesn't develop. Without proper funding, promised biosecurity reforms are unlikely. The LLS today says its aim is 'to build resilient and self-reliant communities that are profitable, productive and sustainable – carried out in healthy, diverse and connected natural environments'. Without the money to do it, what planet are they on?

A Royal Commission into environmental crimes wouldn't be sins committed behind the high walls like religious institutions, or banks, but in full view of Google Earth and along every rural road.

When the New South Wales government shut down the thirteen Catchment Management Authorities (CMAs), and turned them into eleven LLSs in 2014 my political blood pressure peaked. Parliamentarians behave more like interior designers, shoving things around, rearranging the bureaucratic furniture while the farm is on fire, all in the name of efficiency.

It's pointless trying to imagine a perfect government bureaucracy. Overlap and conflicts are inevitable. Think of health issues in schools – they're simultaneously social issues, drug issues, family issues, poverty issues, health issues. The former CMAs had worked with all government agencies and were at the forefront of biodiversity and water management. Most of their staff were plucked from different government departments.

With the CMA transmogrified into the LLS that happened all over again.

Most of us had been enthusiastic about the CMA's charter and were incredulous when yet another bureaucratic whim interrupted the flow of good work. At the very least we'd have to brief our governmental betters – lead our leaders – yet again.

These days I could go to a government-run field day every other day. But cuts and industrial changes mean we've lost the personal relationships of the past, when we could talk easily to a respected cattle officer, an independent well-informed agronomist or native plants expert on the government payroll. Skilled CMA staff were retrenched or put onto contracts. Seeing these talented people drummed out of the regiment was more proof of a lack of respect for local knowledge. We joke sadly that the best business to be in is government printing.

Yes, there were official invitations to apply for board positions on the LLS that would report directly to the Minister of Primary Industry. What I heard from fellow farmers was 'why waste time applying?' and 'the government should just get on with what it should be doing anyway'. RIP CMA. Another headstone for the bureaucratic graveyard. With the LLS unable to match its predecessor's promise, private rural consultancies, many attached to chemical companies, have filled the void.

As a nation we have made some progress managing water but I don't know many farmers who agree. I remember what it was like in the late 1980s before we had state and federal water agreements and state governments were still handing out water licences to anyone who asked.

There can't be a perfect water management system because our personal and business needs always change. Getting furious because a few greedy criminals steal water, lie and cheat is to miss the good news. Laws now declare that the environment is entitled to a drink, and while we're still arguing about how much it should be, changing our attitude of entitlement to water was never going to be easy. Together, we're slowly getting somewhere.

I yearn for the death of 'the market' being considered a masterful arbitrator of how our agricultural water should be used. The flow of money doesn't know 'best use'. If racehorses and wine become the most profitable, is that 'best use'? The market doesn't respond to environmental need, care about our health, business diversity or the importance of a sustainable future. Money flow and water flow remain at odds. Once we realise that, this debate can yield long-term results.

After well-connected recalcitrants were caught stealing vast amounts of water from the Murray–Darling basin there was an investigation. In 2017 we read the damning report that insisted tough compliance and enforcement rules were needed if water theft was to cease. Soon after a Natural Resources Access Regulator was set up. More proof that there's money to be made in agricultural compliance.

We wouldn't need so much regulation if people (the primary producers) both private and corporate did the right thing. But as the recent Royal Commission into the finance sector shows, too many people don't. Appeals on ethical grounds are never enough – for the sake of money, people will do a lot of unethical things if they think they'll get away with it. The firm hand of the law needs to be even firmer.

The issues are complicated with right-versus-wrong blurred. People get genuinely confused. Farmers and irrigators often misunderstand the entitlements granted by their water licences. And outside the Murray–Darling basin rivers, wetlands and aquifers have rules unique to their own catchment plans.

That's on top of the basic federal, state and municipal rules. Of the three tiers of government the third is deemed least important. Yet it's where most things, for good or ill, actually happen and we witness the consequences of 'the market'.

Like all political debates, away from the news the question of water use plays out with full gusto in the quotidian choices we make, and the successes and aching regrets that come from them. With blazing temperatures not everything can survive. Even mighty trees die. Great oaks do not always grow from little acorns. Thirty years ago, I gathered fallen acorns (*Quercus robur*) from a neighbour's majestic tree and felt a genius when they germinated. I planted three in the garden, potted and gifted others to friends. Some stand tall and strong in Gundy today. My three, as uniquely shaped as any three humans, did it tough. One was mowed over when young and didn't make it to adolescence.

The others grew together until this year one died. Both would have, except we decided to water nearby feijoas and accidently left a tap dripping. The moisture seeped into the roots of the oak.

I'd helped bring these three trees into the world, was casual about the death of one, mourned the death of the second and now fear for the third. You must triage in a drought. What animals and plants should get the remaining water? What is going to be worth saving next time a drought puts life on the line?

Culture and emotion play their part, sometimes perversely. We fear eating so many things. Not enough of us eat wild pig, or the carp that our Asian workers think delicious. Would kangaroos, wild pigs and deer be our best food? Three tough animals living, even thriving in the drought. Behind the shearing shed in the thick of the olive grove I count 50 kangaroos resting under the trees. They hardly move when I walk close by. They look at me as if to say. 'So where do you think we should hang out?' The kangaroos, wallabies and wallaroos almost have the place to themselves. The simple fact is they're winning by feasting on our 'locked up' paddocks. For months they get closer and closer until they hop onto the verandah at night waking us up with their thumping and lounge languidly on their elbows on the dry lawn in the morning.

As land managers we calculate the volume of grass available for our stock animals, based on what they will need per day, but the roo count rose so high our number crunching became futile. The grasses were being eaten, not by domesticated animals but the natives and the ferals.

In the late mornings and at twilight you'd watch mobs of roos leaping away from gunshots ricocheting around the hills. Some locals were shooting to kill, others to scare.

By the end of 2018, professional shooters are given government approval to start culling roos across the district, it felt both essential to keep the remaining pastures and to protect the soil but a shameful waste of high-quality protein rotting in barren paddocks.

The 1992 High Court Mabo decision was seen as menacing by many Australians. It wasn't only giant mining or pastoral companies with their millions of acres who were worried. In the suburbs people were incited to fear for the future of their quarter-acre block – just as local farmers dreaded a land claim on their front paddocks. In the latter case there was a hint of collective guilt. Locals knew that Indigenous people had been driven away, sometimes by their own ancestors.

At the very least the right of Traditional Owners had to be considered by the farmers who regarded themselves as 'traditional landowners' – a tradition that began in the eighteenth century rather than tens of thousands of years ago. But Mabo wasn't the only perceived threat for farmers. Mining companies were undermining their farms. Whatever riches lay beneath their property could be claimed by corporate interlopers with court orders. The insatiable curiosity of would-be miners searching for coal beneath their land caused a panic. Whereas the State of the Environment reports had caused no panic at all. Not even much concern.

Crimes with long pasts like *terra nullius*, deaths in custody, the Stolen Generations and sexual abuse in church organisations will be the over-arching narratives of our era, as they should be. In the coming decades, ecological narratives may well follow. Just as big energy companies are now lawyering up to protect themselves from potential court cases holding them responsible for climate change, there will be a reckoning for those responsible for other forms of environmental destruction. While the ancient narrative of environmental destruction suffers compassion fatigue, issues like Mabo gain momentum when seen as a challenge to private property.

Not-for-profit groups, NGOs such as Lock the Gate and other local anti–coal-steam gas and anti-coalmine campaigners, have succeeded in gathering many conservatives to fight the threat. Now farmers were being treated like our First Australians. Many of these new militants would have opposed Mabo. Now they knew what appropriation of land felt like.

'Coalface' isn't a metaphor but a fact. Today, Muswellbrook is the capital city of the vast Upper Hunter Valley coalfields (as opposed to the Lower Hunter, where there are still some remaining, albeit smaller, coalmines among the wineries). Since the first coal was dug at Newcastle not long after white settlement, the mining industry has moved north up the Hunter Valley. It has jumped over the ranges into the Gunnedah Basin and travelled north all though Queensland, extracting gas and coal along the way as towns and ports continue to grow. In February 2019, the Land and Environment Court ruled against a new open-cut coalmine near Gloucester. One of the reasons was that it would impact on Australia's ability to reach its commitment to the Paris climate agreement. Up until this decision, climate change was never considered in coalmine approvals. It brought many of us hope. Yet we must remember that even if no new coal licences were granted, phasing out the fossil fuel sector would be a slow transition, as we wait for long coal and gas export contracts to expire.

Burgeoning alongside the mining boom is the rehabilitation sector. Miners now must factor in rehabilitation during their operations. Although we're yet to see a final mine rehab site worth applauding, often they look not much worse than a badly managed farm. After all, the farming sector has no funds, plans or legal obligations in place to rehabilitate damaged farmland. Agriculture is ironically learning from mining. Farmers visit mine-site open days to witness techniques.

Our renewable energy future comes up against a fight-to-the-death local opposition. We have to ask difficult questions during this transitional time: How can we best rehabilitate the panoramas destroyed by the miners' machines? How can we best restore local agriculture? How can we create new jobs for the dispossessed?

Answers to these questions give new meaning to Muswell-brook's corporate slogan proudly proclaimed on signs as you enter town: MUSWELLBROOK: BURSTING WITH ENERGY.

Experiments with solar techniques, wind turbines and pumped hydro are underway.

In the 1920s Henry Ford assumed his mass-produced cars would be driven by plant-based fuels.[33] Then oil won the day. And here we are again, a hundred years later, going back to the future. A new bioenergy demonstration plant is being built to test ethanol feed stocks. This includes growing renewable feedstock as an economic driver, one that can be decentralised and cross over into agriculture, forestry and technology. Every fossil fuel hub will need to do the same analysis.

Primary production is battling to adjust in the face of the technology 'disruptors', as it tries to keep pace with changing local standards, and the export/import regulations demanded by our trading partners. Protecting the environment in the face of new development has never been harder.

No one knows exactly what post–fossil fuel regional centres will become. But there are plenty of people working, thinking and motivated to prevent these former economic engines becoming derelict and forgotten hell holes.

This coalmining boom won't leave an architectural legacy like the gold rush provided for the likes of Ballarat, Bendigo and Beechworth. While coal built Newcastle, for all the towns north, this latest boom meant money for individuals, not community structures. In a hundred years there'll be little architectural evidence there was a regional coal and gas boom at the end of the twentieth and start of the twenty-first century at all. Except vistas of desolation.

In 2008 a week with one too many farm frustrations had me sign up for the Australian Institute of Company Directors course. It's recommended for anyone entering the corporate sector with hopes of becoming a non-executive director, but is also essential training for people in the not-for-profit sector. Even the organic industry has recommended the course, as important training for small agricultural businesses.

It ran over a few weeks and on the day before we left to study and eventually take the exam, our teacher asked, 'what did you get out of the course?' Each of us had to speak up and then fill out the survey form.

A man in my team who ran a largish company complained that the course 'was interesting but we didn't discuss any of the soft stuff'. The teacher looked perplexed. Soft stuff?

'The fact that people are flawed, liars, sociopaths. That managing such people is crucial. We never discussed that.'

There was silence. Later, in the car park before driving off he returned to the subject.

'We know the rules about corporate governance but can still be blind to the needs of staff, blind to the lunatic in our midst.'

I've often thought about this man. So quiet during the course, refusing to participate in the 'acting out' scenarios, because they were 'just an opportunity for show-offs'. Liars and sociopaths? Could that describe the people stealing water from our river systems? Those whose actions have led the state government to set up a new compliance system? Might the phrase apply to those who so abused migrant workers that we needed to introduce anti-slavery legislation right here in Australia?

One person definitely not a psychopath is Steve the former CEO of our local Waterkeeper Alliance. He announced he was leaving, to open a fresh food shop, a place to buy local food and get a decent coffee in Scone. He too had done the AICD course and previously managed the Australian Stock Horse Association. Now, like me he was entering the food industry.

I was sad the district would lose one of its great educators on H_2O, but, wanting a more creative life he and wife Carey bought one of the many disused shops on Scone's main street, painted it with happy colours and put freshly squeezed juice on a fresh food menu. No big deal? We hadn't had one place in town to get a freshly squeezed juice – so I felt like giving them a fresh squeeze. Thanks to them fresh fruit and vegetables became readily available, chopped up into salads and sandwiches.

In the beginning they sent out leaflets seeking to buy small local vege crops. Were there any? Secretive growers with baggy pants and hoes? We had Morry at Muswellbrook who, in his retirement, grew fresh vegetables beside the New England Highway and sold them from his verandah. And not far away there was an octogenarian who, every spring, sowed heirloom tomato seeds. Turns out few of us were growing vegetables to sell.

During his own domestic harvests Steve would present

freebies. Every winter there was a large bowl of his own limes displayed on the counter.

I turned to Steve when we harvested some garlic scapes one year. These are the leafless flowering stalks that shoot up in the middle of hardneck garlic bulbs before harvest. They can be pulled out of the bulb when green while the garlic is still growing and eaten. I delivered a few bunches to Steve but it was a lost cause, 'No one knows what they are, Patrice, and no one knows how to cook them', he said sadly as most went limp on the shelf.

Proof again that growing something can be the easy part, that even more effort needs to go into educating, selling, promoting. But Steve's other produce sold well with local restaurants who saw advantage in freshness the supermarkets couldn't match and in Steve's skill in putting together their orders. His sons helped in the store during school hols and soon it was a popular place for meeting, greeting, eating.

But it couldn't last. Their clientele, principally people who cooked old-fashioned food from scratch, were getting older. Rural people are as busy as urbanites, and often much poorer. In rural towns the small fresh-fruit-and-vegetable shop is becoming as rare as bookshops and more and more shelves are being filled with 'fresh' processed foods, jams, chutneys, dips etc. The ingredients we need for our health are only available in one, sometimes two stores. There's more choice buying petrol. Steve and Carey's food business closed after a decade.

Most people minding farms today probably won't be there in twenty years. In 2018 the average age of the farmer went down, suggesting a new younger generation is signing up, although none of my friends have their children working alongside them with the intention of taking over the reins. Plenty are contemplating it as they go about their lives studying and working. In the end, family farm succession is as much about inheritance, a fair divvy up of assets, as it is maintaining a business, and now that sisters and brothers alike share a family bounty the idea of equitable distribution is more difficult.

Agricultural success isn't only about the productivity. When a colleague turns her farm into a tourist attraction and makes more money she's a hero of planned diversification.

Non-agricultural diversification options are growing across agricultural land, besides agricultural education – places to touch a pig, milk a cow, Airbnb, open houses, wedding locations, parties, conferences, photography schools, art classes, cooking lessons, camp sites, bike-riding circuits or horse-riding tours can all bring in the cash. Small farms are more likely than ever before to be owner-operator businesses with no staff.

When Jane returns to Scone for her Christmas break she tells me that now in her third year at Agricultural College, over 50 per cent of the class is female and most are from farming families. After graduation Jane 'won't be looking for a manager's job on a private, family-run farm because there's no job growth there.' The most talented graduates, she suggests, want to manage their own family business or become part of a large agribusiness company.

It's a lot harder when you're isolated on a small to medium farm to provide the professional development most young people seek. If an employee can't move up, they move out. A friend in the agricultural human resource industry reminds me how years ago even I interviewed wives when hiring a male manager. Not any more. Wives don't want to be an afterthought to a husband's employment. Qualified male and female managers for small properties are going to be harder to get. There are good people around, but moving to a new rural area isn't only about the employee any more. It's the movement of a whole family. Children's education and employment options for partners are often lacking.

Many cattle operations in Queensland rebuilt their accommodation facilities only to find it didn't really help lure families. No matter how good a pay packet and a house, people can't tolerate the isolation. Location matters more. Some distant areas are destined to be occupied by the unattached. Isolated rural pubs, full of lonely people on bar stools, are a growth sector.

I asked Jane if drought was centre stage in the class room.

'Not really', she said, 'it's part of general risk management.'

'What about drought relief?'

'I don't believe in drought relief,' scoffed Jane.

'Has there been much talk about climate change? Are your courses framed around it now?'

'No, we cover climate change when analysing risk. It's a risk like drought.'

My first lessons on the subject of risk began at school in Kurralta Park, Adelaide, with St Joseph nuns who suffocated in full-length habits, necks and knees never seen. Life was risky, preparedness a virtue, so we'd pray we could manage (we say meditate these days, but it feels the same). We didn't pray for things we personally desired but were guided to focus on those less fortunate managing life's risks, even though we were all a pretty unfortunate lot: plenty of unemployed alcoholic parents, children of refugees who spoke no English at home. At school we were equals. We'd pray silently, kneeling, hands together, fingers pointing to heaven. Though it wasn't the 'done thing' to comment on a particular person less fortunate, on a few occasions such special attention was permitted, as when someone died.

Charity wasn't needed because people were helping everyone anyway. We were all charity cases. Why name the unfortunate? Misfortune touches all of us some time. We were all at risk. And if we were spared – there but for the grace of God go I.

Phillip spent his first ten years growing up on his grandparent's flower farm at East Kew in Melbourne. It was considered run down and shabby by upwardly mobile neighbours. He wore hand-made clothes and was laughed at. He knew the shame of charity and the risk we take when we put our future in the hands of the land.

Purchasing Elmswood was different for me, because I'd never lived on a farm before. I was continuing the long tradition of urban romantics buying rural property. (Since then, real estate prices have risen to such an extent that it's as difficult for young people to buy farmland as it is to get into the city housing market.) The drought, like the nuns, is a clear-eyed teacher. A few years in, a Vinnie's opens up in Scone, fresh competition for the Red Cross. The business of charity is booming.

As farmers, our goal is often a simple one – to do what we love, which is to farm. And consumers want that idea of us, too: the urbanite I used to be wanted to sip a flat white and imagine the organic farmer waking up early, milking the cows, taking time to lean her head against a warm flank. Modern agriculture does not make this vision easy to achieve. Whether farmers are, like Jane, trying to negotiate ever-more-complex international supply chains, or if they've gone the other way and are trying to make it as an independent (our dairy farmer takes a moment to get that perfect Instagram shot), farming is often simply too busy.

And yet, new models are being developed that give farmers the chance to farm and consumers the chance to learn about it, while supporting local farmers. One of my favourites is called community supported agriculture (CSA). Customers sign up, usually for twelve months, to receive a weekly box of fresh farm food.

In California to compare various approaches to farming, I saw many CSA signs on farm gates, and learnt how these farmers were able, once they'd secured their local customer base, to concentrate on their land. And they were having fun, growing unusual produce: multi-coloured tomatoes and differently shaped radishes, cabbages, beetroots and lettuces. In general, customers were still the affluent locals who already love organic produce – but the scheme points to a potential democratisation of the 'paddock to plate' service, through localised food chains that support farmers and nourish their customers.

The system operates with great success, selling not just vegetables but also meats, cheeses and eggs, often with the aid of coolrooms at the back of packing sheds. In Australia CSAs now operate and will hopefully grow – if enough customers can be persuaded to sign up.[34]

Share farming is another alternative that has regained popularity. This system usually has one landowner but profits are shared with other farmers who do the work. Fellow organic farmers have banded together to buy farmland and divvy up responsibilities. These various business collectives share the cost of maintaining houses, sheds, yards, roads, fences and gear.

Even more common is leasing, a cost-effective way to access land without the capital outlay. Many people in our district do it. During 2018 I received over a dozen requests to lease part or all of our farm. The requests came from stock and station agents, neighbours and friends, all needing extra land for their rural ambitions. Agreements vary but many leases run on three-year rolling options. Still, for most the dream is to actually own some acres and, as with city property, access capital gains – and security.

I once thought a farm would be somewhere to hide and get away from the increasing noise of a globalised world. I wasn't alone – and that was before tech companies spread around the world with a business model that deliberately undermines our privacy and individuality, threatening our very ability to be a creative soul. The idea of the farm as a place to escape to holds even more appeal now, but it's delusional. A farm does offer a place for solitary contemplation, but so does your bathroom. And your bathroom is a lot easier to look after.

You don't 'get away from it all' on a block of land. We expect emergency services to find us if we're unfortunate to actually get lost, almost drown or be cut off by a storm or bushfire. Some crises are social, not emergencies. A few years ago, a woman with two children moved into a caravan on the block next door to friends of mine, built a hut alongside it, and then allowed piles of rubbish to collect because she simply couldn't manage. When these piles became too large she would set them on fire, making them go off like bombs. She'd fire guns to amuse herself and play music loudly all through the night. My friends had chosen their half-hectare block for its location in this previously quiet valley. When they tried to ask for more neighbourly consideration, they were screamed at, the music got louder and gunshots closer. The problem escalated as various government departments were notified and struggled to find a resolution. Helping wasn't easy. Eventually my friends moved.

It was a raw example of how people, trying to find refuge in rural hideaways, may not find a place to call home.

(Small blocks on the edge of farmland and near national parks are part of historical subdivisions. They're Lot Numbers, receiving virtually no services, except the occasional road grade. These hidden picturesque blocks are what makes them appealing until isolation becomes abandonment. To resist this trend, we'll need more energy than ever before in building these sustainable communities, where people have the resources to truly support each other.)

Strong communities, like good soil, can produce wonderful things. For farming communities, it could mean turning risk and threat into opportunity.

When Jared Diamond wrote how past civilisations made choices that led to their collapse, we wondered what the conversations were when the last tree was chopped down or died on Easter Island. Were fears expressed when deserts started forming across the Fertile Crescent, that vast arc of land across the Middle East? I imagine the conversations weren't so different from the ones we have here right now about drought or when our local energy company wanted to erect a new substation on a block of land with a salinity problem next to our high school. Isolated small problems never feel like they're part of a bigger problem.

Coverage of 24th Conference of the Parties to the United Nations Framework Convention on Climate Change, or COP24, in Katowice, Poland, during December 2018, reminds us that 500 million small farmers are extremely vulnerable to climate change. Enter climate smart agriculture, CSA (confusingly the same acronym as community supported agriculture). Community supported agriculture is already doing in the developed world what climate smart agriculture is suggesting should be done in the less developed world; provide sustainable, efficient ways to grow and distribute affordable food and manage land. Ideas include silvopasture (where tree crops, pasture and animals are produced together), conservation agriculture, and agroforestry.

To carry out climate smart agriculture, farmers need access to knowledge. Yet in Australia we've closed regional education centres. Historically a TAFE course was proposed to fill in any local training gaps. Now the latest TAFE courses are mostly online – learning in the comfort of your own home/farm is meant to be an advantage. But the internet is rubbish out here, and we're missing a critical opportunity for farmers to meet each other, strengthen community bonds, stimulate new learning and exchange knowledge. We'll need to be tough as old leather, if we're going to get through these next decades together.

It's a particularly hard pill to swallow that the best-funded courses, where students do work together, don't necessarily lead to stronger farming communities but to employment in chemical companies, research, machinery development, IT. There's no record of where all the agriculture graduates end up.

Manufacturers from Volvo to Samsung can control the inputs to their assembly lines because they know the little gizmos needed to make larger gadgets, be it a car or a big-screen TV, will arrive on time and be of the expected size, quality and price. On a farm there are so many uncontrollable variables. We're yet to know how many farmers will leave the sector at the end of this drought and flood era. Or how many farms will be sold, split up or joined together to make different businesses in the future, or how many will go bankrupt.

When we talk drought it's the primary producer who comes to mind, but we know that the tendrils of drought reach far and wide into every part of the community. I try to find out via different DPI staff helping drought-affected people and businesses how many are suffering in our area and get nowhere. It is a privacy issue, I'm told. Not even raw numbers can be given out. Letters to the drought coordinator, our local member and the minister himself go nowhere.

After a report, radio interview, podcast, blog or public discussion on drought, it's become common to mention afterwards how to access government financial services. It's similar to the messaging given after a story on mental health. 'If anything in this report has concerned you, please contact…'

Charles Massy's fine book *Call of the Reed Warbler* describes new-generation farmers choosing a regenerative path for their rural business, using what Massy describes as their activated 'emergent minds', once their old 'mechanical brains' can be laid to rest. He's not talking about brain transplants, but new attitudes he believes we're all capable of learning.

When I started my farm journey I had an 'urban' brain and didn't understand why organic farmers weren't shouting from the roof tops about the menace of Industrial Agriculture. Some of the leading organic farmers of the day told me, it's not up to us to force people to change. People need their own personal light-bulb moment, their activated 'emergent mind,' often triggered by getting sick, losing money, drought. Until then, it's easier to continue with what we know and feel comfortable.

Laws change attitudes slowly. The native vegetation, biodiversity, clean air regulations and the ongoing Murray–Darling basin water wars show how hard it is to reconcile environmental need and the divine right of farmers.

We've been brainwashed into believing a good farm is one that endlessly increases its productive output. We're still only adding up one part of the ledger.

No wonder we're confused. Global statistics and reports from one government department portray problems (Department of Environment) whereas another claims success (Department of Primary Industry).

Australia is classed as a sophisticated agricultural country. Production technologies have helped us export millions of tonnes of grains and animal protein. But the best technology in the world won't supply cheaper water or guarantee its supply.

Between 2012 and 2015, I was part of a group forged by the local Catchment Management Authority and a mighty coalmine. Local mines own a lot of agricultural land around their mine sites and carry out various agricultural activities, mainly beef production but also dairying and hay-making. Our soil-carbon farmer group was trying to tap into the ideas that climate smart agriculture was promoting, especially around increasing soil carbon.

It's the sort of project coalmines fund to prove they have community engagement and are thinking about the post-coal world.

Our strategy was to try anything and everything, to find achievable ways to increase carbon levels in our soils. The group was made up of seventeen farmers across the catchment doing different agricultural activities – hay, sheep, beef, dairy – on very different landscapes. We allocated an area of each of our farms for the project and collectively covered 1000 hectares, and we measured our soils. At the project's end, we measured them again to see if the soil carbon had gone up or down. Every one of us saw some increase, though the levels varied. Some did it through applying inorganic fertilisers, poisoning weeds, adding hybrid seeds to their pasture mix. Some used irrigation, while some, like me, assessed rotational grazing. Results showed a conclusive 1 per cent average increase in soil carbon levels, equal to 15 tonnes per hectare.

We all felt some pride in knowing that our patch could sequester carbon, proof we could be a climate smart agriculture solution. Even if a coalmine next door undid our good works by exporting coal. We worked out the money we could have made if farmers were paid for sequestering carbon, and felt hopeful at the diversity of tactics that had worked. We were under no delusion our 'research' was particularly formal – we were learning the way farmers usually do, getting our hands dirty with trials and failures (though in our case, with slightly more sophisticated measuring equipment). So it should have come as no surprise when, as with so many farmers' tests, a change in the weather called our hope into question. Our research was done between 2012 and 2015, years that turned out to be some of the wettest on record, a golden era for building soil carbon.

Our group lives on under another name, the Upper Hunter Sustainable Farming Group, and field days. In the intervening years, few of us have dramatically changed our agricultural practices. We've been busy learning other things, things we never wanted to learn, about the limits of our skills in the face of a dry riverbed, a dying animal, a business taking its last gasps. But we know that our hotter, drier future will need new experiments.

All of us have wish lists that, when boiled down, are similar. We're in pursuit of a healthier, happier environment, even if the language we use to express it is changing. Once 'transparency' was the word that demanded most attention, then 'sustainability'. Now the vocabulary includes 'pressures', 'trends' 'risks' 'outlook', all part of the 'state' of the environment. Jargon is more prolific than ever before: renewal, rejuvenation, and the ubiquitous resilience.

This book describes many negative aspects embedded in the culture of agriculture today. In my future world, organic food will be more readily available and affordable and we won't be surprised when the next research paper suggests that eating organic food may significantly improve our health.

Joan Lindsay, author of *Picnic at Hanging Rock*, agreed with Einstein that time was an illusion, albeit a particularly convincing one. And as I get older I believe in time less and less. Yet farming is driven by time – not clock time so much as seasonal time. But even that's becoming illusional.

The times, they are a-changing with climate change, not so much a ticking clock as a ticking bomb and farmers are being forced to adjust plantings and harvesting. At the start of 2019 in northern Queensland, floods closed roads, buckled train lines, destroyed houses and businesses, killed half-a-million cattle and forced thousands of people into evacuation centres. At the same time, lightning triggered numerous fires throughout the slowly drying south-west of Tasmania, where 1000-year-old trees and unique ecosystems now face irreparable damage and hectares of leatherwood trees, famous for their nectar, perished. At Elmswood the drought remained. Our emergency has steadily evolved over two years. In Europe and North America, temperatures reached all-time lows.

A garlic grower moved to Tasmania because the long-term weather projections showed decreasing rain and increasing heat for her New South Wales farm. No one is sure where water and kind weather is any more. As a weather watcher, I know that some meteorological prophecies for the Hunter suggest a subtropical future – while others insist that things will get even drier, which seems to be what we're enduring.

Here's the reality: we're not changing fast enough. The 2016 State of Environment report recorded improvements in some built environments. The deteriorating areas, surprise surprise, were in the 'extensive land use zones' (i.e. farms) where grazing is noted as the major threat to biodiversity. These are the unforeseen consequences of 'common practice', of unquestioned 'traditions'. Among the worst are over grazing, poisoning weeds near creeks, clearing along rivers, burning all tree debris, and cutting down old trees for firewood. Stealing water when you can get away with it.

Prior to European settlement, ecologically diverse and fragile native grasslands were everywhere. In Victoria only 1 per cent of the natural temperate grassland remains intact today. There are too many old wounds to heal and must to come to terms with the sad fact that much damage will not be fully repaired. Legacy issues, like the greenhouse gases in the atmosphere, missing plants in the landscape, erosion, water diversions.

Behind this destruction was a universal view that native grasslands were of little use. I still hear that today. 'Bush' is a pejorative; grasses need to be 'improved' for productivity. I can comprehend farmers seeing native grasses as less productive – they're bombarded with puffery regarding fertilisers and new hybrid seed stock. But I cannot grasp their failure to see native grasslands as beautiful.

Once, when Aurora was little, we had a massive rainfall in early spring and native flowers popped up overnight. It was sudden and exciting. I took her and a schoolfriend out to see this overnight transformation and asked them to pose for a photo amidst its abundance. I found the photos a few days back, in cruel contrast to the drought. The next time my grown-up daughter is back, I'll show her. Will she even remember?

Stories of the family farmer soften the image of Big Ag. The image of people farming tug at the heart every time there's a drama – drought, typhoon or flood – when crops and people's lives are ruined. At these times city people do want to help because deep down they believe in the farmer mythology. Kids have learnt it from singing old MacDonald had a farm E-I-E-I-O. It's all mooing, oinking, neighing, cluck-clucking, baa-baaing and quacking away, becoming a part of our kinder-garten memories. But these 'family farmers' are a vision of tradition that can be dangerous. I want our collective imagi-nation to be something more: a story of communities working together to overcome problems, not just emergencies. Of creative and determined people finding fast, workable solutions.

Of people who care for each other, not just directly, but also by caring for the land. I think we can get there. But we're not there yet.

We all want work that offers satisfaction. 'Find the job that you love, do what inspires you.' And I found mine. No job in the world offers only good times. Having grown up with parents who worked only for money and never had a job they actually liked, I was determined not to follow in their footsteps. Though their life wasn't miserable, it was (to me) boring. On a farm most work is dirty work, hard work, endless. Yet it's varied, with no two days ever quite the same. And our under-the-sky office can be beautiful.

The truth is, after three decades, I expected to achieve more environmental improvements to this land I farm. In good times we've made real leaps at Elmswood, but I am not the master of nature. I can help it, but also easily hurt it. And my time frame isn't its time frame. What the farm has taught me is to be patient. And to realise that despite being an eager student of nature, sometimes I don't remember the lessons.

My actions are part of the future of farming. Every farmer and potential farmer is part of that future, whether they're old farmers ready to respond to the climate emergency with new ideas, or bright young things from Environmental Studies, coming to the land to apply their principles. Every decision each of us makes shapes what comes next. We've lived through the social justice, feminist and civil rights eras. We know the importance of speaking up. That's what we do these days. Kids do it. My dogs do it. 'I need

a cuddle' CJ thinks, and immediately her head is on my thigh and she gazes lovingly at me to pat her. She's not thinking, 'should I let Patrice have her cup of tea first?' Never. She knows her needs.

The land is actually like that. It's asking but we don't hear. Land does speak to us without words. We have to listen to knows its needs.

Much like my progress with Elmswood's ecology, the progress for organic farming (or the organics industry) in the last decades has been erratic. In 2015, I signed up for the Australian Organics tour to the USA, to visit organic farms and businesses across California, as well as the Natural Products Expo at the Anaheim Convention Centre.

'Natural' turned out to be more fantasy than fact, more at home in nearby Disneyland. So many deceptive labels and attempts to flog processed foods as God's gift. But there were still plenty of dinky-di organic people to passionately discuss sustainability and show-and-tell us about improved packaging to tackle the curse of one-off plastic bubble packs.

Things were more natural when you escaped the main halls

and went to the booths out the back, which were far cheaper to rent. Here were intriguing displays of plants from South America and Africa I'd never heard of. Exhibitors were competing to convince us that their fruit or vegetable would be the world's next super food.

Some of the exotics I saw may become as familiar as bananas. Or blueberries which, when I was a kid, seemed as remote and American as a trip to Disneyland. Instead we had blackberries – escapees from gardens gone feral along the roadsides in the Adelaide Hills, with thorns to tear at the flesh and fruit to stain the fingers. I didn't taste a blueberry until I was 24 and living in New York

The commercial operator making food does not have the consumer's health as a priority. Food manufacturing pursues cheap ingredients. As I walked from stand to stand, my ID badge showing my name, farm and a barcode, people asked me what 'ingredient' I produced. I had become their ingredient producer. They'd use their phone to scan my badge, hand me a card and later that day send a promotional email.

The global economy provides increasingly novel foods for the rich and indulgent. Trudging up and down the Expo's aisles I squirmed at the mountains of 'brand new processed organic foods'. Dried bananas dipped in peanut butter. Delicious. Rose-flavoured kombucha soft drinks. Thirst quenching. I'd been brewing kombucha (a drink that's been made for centuries by fermenting black tea with sugar) and making vinegar out of leftover red wine ever since discovering Bill Mollison's book on fermented foods. Kombucha's 'mother', called a 'scoby' looks awful, better suited to a laboratory than the kitchen. But ah, the taste of the tea! Now kombucha's being marketed like Coca-Cola, as a range of 'new' and miraculous health drinks, with pressure-cooked carbonisation, loads of extra sugar and fake floral flavours.

A wide variety of powdered food supplements were also being flogged. Dried and ground organic vegetables to sprinkle on your yoghurt or pep up your smoothies. Was this the future of organics? I felt queasy, uneasy. Already many of Australia's organic dairies were having their fresh milk dried into a powder to be used in 'natural' processed foods.

And yet it was good to learn that most dairies in northern California are now organic, with small creameries making a variety of specialist yoghurts, ice-creams, cheeses and drinks. Cows eating actual grass in one paddock, and milk products being made in little creameries beside the milking sheds. We visited one of these new-age milk producers, hidden away in the hills beyond San Francisco. It felt like home, with poor roads, run-down fences, reminiscent of the central north coast around Comboyne, the hills around Maleny or the gentle slopes around Warrnambool. Remember Cadbury's claim of chocolate being made of the milk from contented cows? In places like this – with these San Franciscan Friesians – the cattle really are content.

Back in town we visit the fabled Chez Panisse, created by

Alice Waters in 1971 to demonstrate the principle of 'paddock to plate'. Alice stops by our table to meet her Aussie comrades. We eat far too much but feel entirely virtuous. Alice started something when she opened that restaurant that has been felt in restaurant kitchens around the world. Even fast-food chains use some of the language she popularised. It's an uneasy victory, but a victory nonetheless.

Look closely, and you see changes everywhere. In 1980, I was living in a New York loft in downtown Tribeca, in Worth Street. Revisiting this nondescript street in 2018 I discovered a revolution under way in a ramshackle building I used to pass every day. Not a political revolution, but a food revolution. After a walk down a bleak hallway I was welcomed to, of all things, a miniature farm, the size of an average lounge room. A hydroponic farm. The only clue to its existence was an easy-to-miss street level sign reading Farm.One. Here was a new type of fast food. A variety of plants grown very quickly using LED lights and New York's tap water. There were leafy greens ranging from the exotic to the humble lettuce. A lettuce that would take six weeks to reach maturity in the soil was ready to have its leaves plucked in one.

If you see agriculture as just business then food production in a laboratory is an easy shift. New farm technologies in a new space. But if land is your thing, the ideas of the future are more complex.

Even though we all need food, growth in demand for food compared with other services is slowing. The developed world is spending a lot more on entertainment and travel than food.[35]

As an organic farmer I'm obsessed with soil. Caring for its complex biology is my life. I believe that healthy soil is the foundation of wholesome food. But there in the middle of Manhattan, this view was being challenged by the heresy of hydroponics. The idea of growing food without soil raises my eyebrows, hackles and loads of questions.

I was introduced to Farm.One by Misha Hyman, a celebrity chef and self-described 'food warrior', one of the chef/fresh food/health advocates, billed as an 'influencer'. He agrees with old-fashioned soil farmers that good food is the best medicine, but equally important is quick delivery of fresh produce. Of food grown upstairs in a mysterious fluid and artificial light? More and more New Yorkers and restaurants are liking the idea – a few days later I discovered another hydroponic 'farm' in the centre of Grand Central Station, serving hungry travellers in an upscale outlet. The basil was growing in something that looked like a huge fridge, next to the pizza oven; it was clipped and then served. Harvest to plate: five seconds. Transport logistics: five steps.

Some environmental battles are fought over and over again, like the fight against each and every new coalmine or coal-seam gas flare. Even for the participants it can be strangely boring – the same words, same warriors, same weapons. Bringing the battle to the cities, where most Australians live, can surely only be a good thing. We need to stop thinking of the environment as being 'out there'.

Imagine if every apartment block in Sydney or Melbourne

had a farm inside. New buildings often have gyms and swimming pools to help the health of residents. So why not farms? In the future will the right to farm include the right to put a hydroponic farm in your spare bedroom? Think what all the empty nesters could do with their under-utilised space. With farming moving into apartments, the old farmer versus environmentalist battle-grounds would shift. Our vantage point reframed.

Speaking at the 2018 Soil Conference in Canberra, soil scientist Professor Alex McBratney suggested relying on what he calls our 'thin skin of soil' for the food of the future may not be wise. Synthetic biology could be the answer with vertical farms, hydroponics, glasshouses if we can make the massive infrastructure investment necessary.[36] Removing our reliance on soil will free up the wounded soils and give them a chance to heal.

An essential challenge for garlic growers everywhere is that modern cooks seem to find using a fresh garlic clove too much work. Far easier to open a jar. Alongside the fresh garlic I saw in the USA were dozens of jars and packets of crushed, powdered, sliced, smoked and flavoured garlics. The variations in flavour, presentation and price were impressive but the main claim was the new essential, convenience. If it isn't the cheapest product it has to be the most convenient.

Give a customer what they want goes the saying. But isn't it better to explain why a fresh bulb of garlic is an essential kitchen ingredient, a medicine, and above all an opportunity to be creative? When you use fresh ingredients to cook dinner, the end result is a little different every time, because no two vegetables are ever exactly the same. If a recipe calls for one clove of garlic, the question is: one clove of what, exactly? A clove of just-harvested, uncured garlic is softer. A clove of Chinese imported garlic can be bland. It's easy to add too much garlic if you're not careful.

My daughter tells of cooking garlic custard for her in-laws a few days after meeting them. At home, it's a family staple and we use a full bulb; the Californian variety she was using was not so mild and turned the normally sweet garlic custard dessert into a savoury dish. From a vegetable seller's point of view, you can see why varieties just disappear. We need buyers to know and appreciate differences.

That first garlic crop we planted was on a little stretch of land not much bigger than a cricket pitch. Each year it colonised more space, until, at its peak, before the latest drought, it was the size of a cricket oval, and threatening to leap the fence and take over the auditorium. Graeme was there right from the beginning, the Test captain, an inspiration to me and the team.

For years, he came and went, with Betty, John and all the casuals working alongside him, most of them a quarter his age, although that didn't stop him competing with every one of them. Graeme was always alert and observant, vital not just to our garlic but to our whole farm. But he also belonged to others and other parts of the country. He would visit each of us for months at a time, driving in great loops around New South Wales and Victoria.

Then Graeme went missing. He hadn't turned up when we were expecting him to, and when he didn't answer my texts I phoned his scattered family, to learn that he was in hospital after a mild stroke. At the age of 67 he'd never before been a patient. He confessed to enjoying the hospital food, the nurses and the fuss. The big problem was he'd have to give up smoking and make do with nicotine patches. No more roll-your-owns made from the illegal tobacco he bought from mates.

Three years on, we threw a surprise birthday party for his seventieth, which reunited his fractured and scattered family. We heard again his ancient jokes and anecdotes about the years he'd spent driving huge trucks across the Nullarbor. Now he was reduced to driving heavy tractors across our paddocks.

Finally, after an especially good garlic crop, Graeme chose to end his farming. It was a time of sorrow and celebration at Elmswood. He drove out the gate for the last time with his car loaded, to head overseas and spend his retirement and his pension. But it wasn't to be for long. A few months later we heard of his tragic death. He had survived being gunned down one night on a Cebu footpath in the Philippines. A few days later, he chose to tear himself free from the tubes and wires keeping him barely alive in his hospital bed.

When social media took off, the agriculture sector saw a plethora of small grants going to people and organisations to develop agricultural apps. Networking is important and it amazes me how many locals regularly visit the Scone Community Facebook page, or the Scone Buy Swap group. My phone has been full of apps meant to make me a better farmer. I've deleted many of them, keeping the essentials: BOM, NSW Fires Near Us, and Field Guides. It will be a sad day if I remove from the glove box my books on birds or weed identification.

And the apps can only work if you have decent mobile reception. Which we don't. Instead I rely on three methods of meteorological prediction, a combination of nature and technology. The first two are frogs and dogs. Should there be any hope of rain, the green tree frogs on the verandah begin croaking. One seems to live in a down-pipe, so its efforts are echo-chambered. When a storm is on approach our dogs, Squire and CJ, hear the thunder long, long before it's audible to humans and slink away to cower in the laundry.

The third source of information is the Bureau of Meteorology, or as we say, the BOM. It provides weather obsessives a lot more than rain forecasts. It tells of wind direction and also measures the flow of water through our landscape, first by measuring any rainfall entering a grid cell and then, as it moves as run-off, into drainage, ground water or dams. Or is lost to evapotranspiration.

It's telling us what we already know but the app helps me face the facts of drought. At the start of 2019 BOM data continually reported soil moisture in steady decline, predicting a lack of water in our river, dams and, most importantly, in our soil. After so many floods it was hard to remember sub-soil moisture as a problem. After a small shower, the soil dries almost immediately.

The water just disappears into the magnet of deep dryness. More alarmingly the BOM reveals that despite wet seasons during the decade there's been a twenty-year decline in soil moisture right across south-eastern Australia.

The BOM's 'root zone soil moisture maps' show more than a drought. Summers start and keep going right into autumn. Relentless, unrelieved. No fun for the frogs, no frights for the dogs. I can't help looking constantly at my phone in the hope of good news, getting only the harsh reality of another 30-degree day. It was hot yesterday and the day before and it will be hot, even hotter tomorrow. On and on and on. You think you're hot? How do you think your soil feels?

We distract ourselves with dinners sitting down to barbecued lamb chops, bunches of fresh chopped herbs and our new favourite, grated long radish salad, tossed in a dressing of olive oil, honey, apple cider vinegar, fresh herbs and toasted pecans. Dinner discussions are framed by drought, and we mention local wedding plans with so little water available to green-up the place and control the dust. To an open garden event, where water is being poured onto thirsty introduced plants. Feelings are polarised as we debate whether a green oasis offers respite or is it a total waste of water for a short-lived floral fantasy. Every choice we make is heavy.

Speaking about agriculture and therefore the environment truthfully has never been harder. All of us involved in rural campaigns, irrespective of category, run the same arguments, meet the same politicians, and often we fail. Like I failed when I spoke to a neighbour about irrigation. We need to change strategies, tactics, and develop new perspectives.

For example, if we consider the issue of land clearing, environmentalists blame farmers.[37] And in return farmers blame environmentalists for bushfires that they insist are worse than when land isn't cleared. Out of all the environmental issues that have erupted since I moved to the farm, the biggest sticking point has been that generally farmers believe farmers know best.

Prime Minister Julia Gillard spoke of the importance of nurturing your sense of self. She said resilience (there's that word again) might be the main requirement for women if they enter the cesspit of parliament or any part of civil society. Nurturing is the foundation of that resilience. Julia's misogyny speech now has more than 3 million YouTube views (I required some self-nurturing after re-watching it), and Gilliard clearly drew on hers to deliver it.

A few years ago, a local twenty-four-year-old killed himself on Christmas Day. He was a young football hero, a successful electrician living in our community with his extended family. Yet on a day dedicated to happiness, he ended his life.

The numbing news passed around the district. In an effort to make sense of it, the young man's family needed to do something. They'd had no idea he was suicidal and asked the question: Why? The family decided to encourage a local conversation about suicide and soon a not-for-profit called Where There's a

Will was born in his honour. Will was the name of the young man who died.

'Where there's a will', is often said without the rest of the saying. Because we know it without saying it: there's a way. The team campaigns for a positive psychology education program, a mental health first-aid course, to be rolled out across the Hunter Valley. Funded by private donations, the program is available to everyone. Pauline, Will's mother, who spoke to her son the day before he died, often repeats a particular statistic: a young man is three times more likely to die from taking his own life than in a car accident. Still we worry so much more about road deaths.

The powerful message from Where There's a Will is about caring, observation and communication. And it's told here in the area, face to face. Not just via the internet, not only heard on the radio, or watched on television but between neighbours, shop assistants, teachers, friends and families. It's not gender-based. It's people-based. Most teachers in the area have now been trained and apply it in the local schools. Special classes are given to primary school students.

A fellow farmer, a woman, says to me with weary frustration. 'Nothing will get better in agriculture until farmers' mental health is fixed!' So it's exciting that our local Where There's a Will program is helping the next generation of our community to be better mentally equipped.

It's a campaign that gives me hope. And I'd like the principles applied to the land as well. There are profound and proven links between mental health and environ*mental* health. No wonder the terms are so alike. But all the farm courses in the world won't do a thing until deep down everyone who manages land understands that relationship and refocuses their work to nurture both people and land.

On receiving the Jerusalem Prize for the Freedom of the Individual in 2001, Susan Sontag made the observation that what the writer writes isn't so important. It is what the writer is. This compelling idea has, I believe, universal application and glaring examples. It is not what the priest says, it is what the priest is. It is not what the politician says, it is what he or she is. It is not what the farmer says, but what the farmer is. So too the environmentalist.

Similar thinking had Judith Wright move away from poetry to environmental activism. She said we needed environmentalists more than poets. I'm not sure I agree. Wright wrote with reverence about the natural world, it imbued her work and drove her idealism. Younger poets continue the ancient tradition of nature poetry, using different words, rhythms, and metaphors, helping us see and hear afresh. If you want to deepen your reverence, become friends with a poet. Or read a poem. Or better still, write one. But do it to improve who you are, not for the writing itself.

Nothing will work unless you do, says poet Maya Angelou. She's right, and there's a lot to mind. We'll need an army of minders to do it. The farmers, the blockies, the land managers, the glasshouse operators, the urban gardeners, the fun lovers, the eaters. Nothing will work unless we do. All of us.

ACKNOWLEDGEMENTS

WHEN I STARTED THINKING ABOUT THIS BOOK, THINGS WERE just starting to get a little thirsty at the farm.

Without my traditional distraction of gardening, impossible without water, I turned to music, visitors, voices on the phone, faces via Skype. The need for connection grew stronger as the drought became ever harsher.

The sensible response to a drought would be to hide from all the non-events – the non-rain, the non-growing of crops. Finally, there was no choice. A book had to be written. Help arrived in a thousand different ways and frequently beyond the call of duty.

Firstly, thank you Ben Ball for originally commissioning the book and editor Meredith Rose who helped shape it from notes and provided essential tough comments along the way. Special thanks to publisher Meredith Curnow for her steady counsel and managing editor Clive Hebard who patiently put it together, Alex Ross for the cover design, Susan Keogh for proof reading and publicist Bella Arnott-Hoare.

Like farming, publishing is a team effort.

Hundreds of people have helped grow the food and minded the farm while I wrote. In no special order, blessings to Betty Chamberlain, Richard Ali, Jon May Steers, Dean Taylor, Jonathan Randle, Jane Pinkerton, Buck Emerton, Geoff Lloyd, Steve Tilse, Iain Hayes, Caroline Hayes, Pip Baker, Sue Abbott, Jill Reid, Kim Jenkins, Anita Lawrence, Linda Russell, Ivy Muffett, Anne Fox, Fiona Marshall, Judy Daunt, Lindsay Jerrick, Newell Lock, Professor Alex McBratney, Emeritus Professor Tim Roberts, Dr Steven Lucas, Emeritus Professor John Rodger, Dr Annelie Wenslandt, Dr Andrew Monk, Kerrie Burns, Steve Osborn, Danielle Lloyd Pritchard, Alec Roberts, Phil Gilbert, John Gilbert, Sophie Frazer, Tobias Keonig, Beatrice Koenig, Mark Saywell, Rosie Barr, Peter Bennetto, Melanie Gillet, Suzie Ward, Gideon Warhaft, Eli Brassé. Fellow biodynamic and organic farmers. My neighbours near and far. The wwoofer office team and all the wwoofers who've helped us grow food at Elmswood.

My gratitude to Roger Sternhell in Sydney who doggedly helps with all the retailing. Artist Katherine Smith for creative attention to all our artwork. Lindsay Hodge, Virginia Reynolds and the staff at Scone Post Office, and the many others at Australia Post.

Thanks to John Sullivan who now lives at Elmswood with his family – Tarah, Tyrohne, and Shaelee – as we prepare for the next post-drought era. Special love to Aurora Adams and Susannah Compton who continue to love Elmswood. And Phillip who has shared it all.

Notes

Introduction

1

W. Steffen, et al. (2011). 'The Anthropocene: from global change to planetary stewardship.' *AMBIO: A Journal of the Human Environment* 40.7: 739.

W. Steffen, P. J. Crutzen & J. R. McNeill (2007). 'The Anthropocene: are humans now overwhelming the great forces of nature?' *AMBIO* 36.8: 614-622.

2

For more information of why we need a world soil day on 5 December, see www.fao.org/world-soil-day

3

www.ifoam.bio

4

The word chemical is complex. Yes, we are all made of chemicals. The world can be reduced to chemical elements and molecules, but I am not a chemist and don't use the word organic (carbon containing) and inorganic (non-carbon containing) as they would.

5

When using the term certified organic agricultural systems, this includes farming as a production practice, processing, transport and packaging. To become a certified organic or biodynamic supplier the regulations are many and constantly being updated to match overseas standards. To read in full the ACO Certification Ltd requirements per industry, see www.aco.net.au/

Australian Organic Market reports are conducted each year by the industry and provide the best up-to-date statistics. See: https://austorganic.com/wp-content/uploads/2018/04/AustOrganicMarketReport2018_spreads_digital.pdf

6

Carbon dioxide equivalent, CO_2-e, provides the basis for comparing the warming effect of different greenhouse gases. For updated information on Australia's greenhouse gas emissions, see www.environment.gov.au/climate-change

L. B. Guo, & R. M. Gifford, (2002). 'Soil carbon stocks and land use change: a meta-analysis.' *Global Change Biology* 8(4), 345-360.

CHAPTER I
7

S. Ankri, & D. Mirelman (1999). 'Antimicrobial properties of allicin from garlic.' *Microbes and Infection* 1.2: 125-129.

R. S. Rivlin (2001). 'Historical perspective on the use of garlic.' *The Journal of Nutrition* 131.3: 951S-954S.

R. S. Rivlin (2006). 'Is garlic alternative medicine?' *The Journal of Nutrition* 136.3: 713S-715S.

E. Block (20110). *Garlic and Other Alliums: The Lore and the Science.* Cambridge: RSC Publishing.

8

G. M. Volk & D. Stern (2009). 'Phenotypic characteristics of ten garlic cultivars grown at different North American locations.' *HortScience* 44(5), 1238-1247.

Around the world, the appearance of garlic – its phenotypic plasticity, as scientists call it – varies greatly. Garlic cultivars or classes grown under diverse conditions have highly elastic soil nutrient responses, particularly relating to skin colour and yield.

9

K. Ried et al. (2008). 'Effect of garlic on blood pressure: a systematic review and meta-analysis.' *BMC Cardiovascular Disorders* 8.1: 13.

This meta-analysis of garlic preparations showed it was superior to placebo in lowering blood pressure.

K. Ried & P. Fakler (2014). 'Potential of garlic (*Allium sativum*) in lowering high blood pressure: mechanisms of action and clinical relevance.' *Integrated Blood Pressure Control* 7: 71.

This research reviewed how aged garlic can help lower blood pressure. It showed a reduction in blood pressure of about 10 mmHg systolic and 8 mmHg diastolic, similar to standard BP medication. Their report reviews many of the historical uses of medical garlic and the references are most useful. See, www.ncbi. nlm.nih.gov/pmc/articles/PMC4266250/

Chapter 4

10

Biochar in soil. For general information about how charcoal was used to build ancient civilisations and how we can reintroduce some techniques, three general books stand out.

J. Bruges (2009). *The Biochar Debate. Charcoal's potential to reverse Climate Change and Build Soil Fertility.* White River Junction: Chelsea Green Publishing.

A. Bates (2010). *The Biochar Solution.* Gabriola Island: New Society Publishers.

P. Taylor, editor (2010). *The Biochar Revolution. Transforming Agriculture and Environment.* Mt Evelyn: Global Publishing Group.

11

https://reneweconomy.com.au/origin-says-solar-cheaper-than-coal-moving-on-from-base-load-70999/

12

For more technical details on slow pyrolysis and biochar see:

J. Lehmann & S. Joseph (2009a). *Biochar for environmental management: science and technology.* London: Earthscan.

To keep up to date with biochar research and projects, see the International Biochar Institute (IBI): https://biochar-international.org

13

To find out more about the commercial slow pyrolysis plant at Tantanoola in South Australia, see the manufacturer's site: www.rainbowbeeeater.com.au

14

P. Newell (2015). 'A strategic assessment of the potential for a new pyrolysis industry in the Hunter Valley' is available at the University of Newcastle's online repository: https://nova.newcastle.edu.au/vital/access/%20/manager/Repository/uon:20006?view=list&f0=sm_creator%3A%22Newell%2C+Patrice%22&sort=ss_dateNormalized+asc%2Csort_ss_title+asc

15

In 2018 a new research group was set up to analyse the truth about Australia's food waste problem. The Fight Food Waste Cooperative Research Centre's first report claims Australia wastes 40 per cent of the food it produces, and individual households throw away around $4000 worth of unused food per year.

CHAPTER 5

16

For more details on domestic violence, see the Australian Government's Australian Institute of Health and Welfare report 'Family, domestic and sexual violence in Australia, 2018'.

https://www.aihw.gov.au/reports/domestic-violence/family-domestic-sexual-violence-in-australia-2018/contents/summary

CHAPTER 7

17

The Australian Bureau of Statistics collects agricultural data every five years via the Agricultural Census. The last one was 2015–16. See 'Australian Bureau of Statistics. Labour Force, Australia, Detailed Quarterly May 2017.' Catalogue number 6291055003.

18

Allan Savory's original classic – A. Savory (1999). *Holistic management. A new framework for decision making.* Washington, D.C.: Island Press – was updated in 2017 with his partner Jody Butterfield: A. Savory & J. Butterfield (2017).

Holistic Management A Commonsense Revolution to Restore Our Environment.
Washington, D.C.: Island Press. Both books are inspired reading and relevant for
all business.

Savory's 2013 TED Talk (now with more than half a million views) places his
work within our climate emergency: www.ted.com/speakers/allan_savory

19

Cattle in feedlots are fed a mixture of high-protein grains, roughage (such as nut
shells and cottonseed meal because their diet is around 60 per cent fibre) vitamins
and antibiotics. According to a Virginia Tech research team, manure from cattle
administered antibiotics drastically changes the bacterial and fungal make-up of
surrounding soil, leading to ecosystem dysfunction.

https://phys.org/news/2017-03-cattle-antibiotics-disturb-soil-ecosystems.
html

D. Zhu et al. (2018). 'Antibiotics disturb the microbiome and increase
the incidence of resistance genes in the gut of a common soil collembolan.'
Environmental Science & Technology 52.5: 3081-3090.

L. Du & L. Wenke (2012). 'Occurrence, fate, and ecotoxicity of antibiotics
in agro-ecosystems. A review.' *Agronomy for Sustainable Development* 32.2 (2012):
309-327.

20

Extreme weather events have repeatedly negatively impacted on intensive animal
production. In 2018, effluent from North Carolina's pig farms spread far and
wide after flooding from a hurricane.

www.nytimes.com/2018/09/19/climate/florence-hog-farms.html

www.newyorker.com/news/news-desk/could-smithfield-foods-have-
prevented-the-rivers-of-hog-waste-in-north-carolina-after-florence

S. J. Khan, et al. (2008). 'Chemical contaminants in feedlot wastes: concen-
trations, effects and attenuation.' *Environment International* 34.6: 839-859.

M. Klein et al. (2010). 'Monitoring bacterial indicators and pathogens in
cattle feedlot waste by real-time PCR.' *Water Research* 44.5: 1381-1388.

J. Shen et al. (2016). 'Ammonia deposition in the neighbourhood of an
intensive cattle feedlot in Victoria, Australia.' *Scientific Reports* 6: 32793.

21

Wohlleben, P. (2016). *The Hidden Life of Trees: What They Feel, How They
Communicate.* Melbourne: Black Inc.

22

Efforts to establish large-scale European style agriculture in the Amazon basin failed because of the nutrient poverty of the soil. Amazonian forests with their huge biomass have a unique ecosystems function.

B. Glaser et al. (2001). 'The "Terra Preta" phenomenon: a model for sustainable agriculture in the humid tropics.' *Naturwissenschaften* 88.1: 37-41.

W. Sombroek et al. (2002). 'Terra preta and terra mulata: pre-columbian Amazon Kitchen middens and agricultural fields, their sustainability and their replication.' *Embrapa Amazônia Oriental-Artigo em anais de congresso (ALICE)*. In 'Symposium Anthropogenic Factors of Soil Formation, 18.; World Congress of Soil Science', 17. Bangkok. Trabalho apresentado. Bangkok.

J. Major et al. (2005). 'Weed composition and cover after three years of soil fertility management in the central Brazilian Amazon: compost, fertilizer, manure and charcoal applications.' *Weed Biology and Management* 5.2: 69-76.

S. P. Sohi (2012). 'Carbon storage with benefits.' *Science* 338.6110: 1034-1035.

23

For the latest details on our climate change emission reductions, see Australia's National Greenhouse Accounts. 'Quarterly Update of Australia's National Greenhouse Gas Inventory: June 2018 Incorporating emissions from the NEM up to September 2018': www.environment.gov.au/system/files/resources/e2b0a880-74b9-436b-9ddd-941a74d81fad/files/nggi-quarterly-update-june-2018.pdf

For up-to-date information easily explained on all aspects of climate change, see the not for profit Climate Council: www.climatecouncil.org.au/

U. Stockmann, M. A. Adams, J. W. Crawford, D. J. Field, N. Henakaarchchi, M. Jenkins & I. Wheeler (2013). 'The knowns, known unknowns and unknowns of sequestration of soil organic carbon.' *Agriculture, Ecosystems & Environment*, 164, 80-99.

24

B. Pascoe (2014). *Dark Emu Black Seeds: Aboriginal Australia and the Birth of Agriculture*. Broome: Magabala Books.

25

To see where your tax dollars are spent in the emissions reduction fund visit their website: www.cleanenergyregulator.gov.au/ERF. The projects the government accepts to fund are updated after each auction round. The eighth Emissions

Reduction Fund auction was held on Monday 10 and Tuesday 11 December 2018.

26

The Federal Government's Australian Institute of Health and Welfare tracks our eating habits. See www.aihw.gov.au/reports-data/behaviours-risk-factors/food-nutrition/overview

In 2017 CSIRO reported Australians don't eat enough fruit and vegetables for a healthy diet. See www.csiro.au/en/News/News-releases/2017/Diets-Lacking-in-Fruit-and-Vegetables

Chapter 8

27

In 1997 the Minister for Health and Family Services and the Minister for Primary Industries and Energy established the Joint Expert Technical Advisory Committee on Antibiotic Resistance (JETACAR). The committee broadly assessed the use of antibiotics in food-producing animals, the occurrence of antibiotic resistance and its importance in human and veterinary medicine. In September 1999 it published a report entitled 'The use of antibiotics in food producing animals: antibiotic-resistant bacteria in animals and humans'. The report concerned:

- the emergence of resistant bacteria in humans and animals following antibiotic use;
- the spread of resistant animal bacteria to humans;
- the transfer of antibiotic-resistance genes from animal bacteria to human pathogens; and
- the possible emergence of resistant strains of animal bacteria which may cause human disease.

(See The Commonwealth Government response to the report of the Joint Expert Technical Advisory Committee on Antibiotic Resistance (JETACAR), August 2000).

There are many feed-mix brands on the market for all animals. Most are supplemented with vitamins and minerals, many include antibiotics.

Because acidosis is a health problem in cattle when they are fed high-grain diets – as in feedlots – the feed is often mixed with antibiotics.

See https://apvma.gov.au for a full list of all the different pesticides and drugs approved for veterinary use.

Rumensin is an antibiotic used as an animal feed additive for cattle (beef and dairy), sheep, chickens and goats and is sold pre-mixed with the feed. Usually 100g is added to a 25kg bag of feed. It's promoted to improve feed efficiency thus help weight gains, control bloat (a problem when gases form in the rumen causing death), increase milk production and improve reproductive performance.

Salinomycin sodium is added to increase growth rates and help feed conversion.

Virginiamycin – a streptogramin class of antibiotics – is added to help an animal eat their high feedlot carbohydrate diet. APVMA publish their findings and the 2004 review of the antibiotic virginiamycin can be found at https://apvma.gov.au/sites/default/files/publication/14231-virginiamycin-final-review-report.pdf

As another example of how Australian and overseas markets treat drugs and chemicals differently, virginiamycin is not used in animal feed in Europe. This is what the APVMA have to say:

1.3.1 Regulatory status in the European Union
'In the European Union, virginiamycin was originally authorised as a feed additive for growth promotant purposes in pigs and poultry. In 1998 the Council of the European Union withdrew the authorisation for the in-feed growth-promotant use of several antibiotics including virginiamycin. This regulation did not affect any prophylactic or therapeutic uses of antibiotics in food animals. However, virginiamycin is not authorised for such uses in food animal species in Europe and does not have an established maximum residue limit (MRL). The decision to withdraw the growth-promotant use of virginiamycin was made despite advice from the council's scientific advisory committee that there was insufficient evidence regarding the transfer of bacterial resistance from livestock to humans. Pfizer Animal Health SA, as the only producer of virginiamycin in the world, challenged the council's decision in the European courts. However in 2002 the Court of Justice upheld the decision, concluding that despite uncertainty as to whether there was a link between the use of antibiotic additives and increased resistance to those antibiotics in humans, the withdrawal of authorisation for the products was not a disproportionate measure given the need to protect public health.'

28

The World Health Organisation International Agency for Research on Cancer (IARC)'s 'IARC Monographs volume 112: valuation of five organophosphate insecticides and herbicides' (20 March 2015) says of glyphosate:

> Glyphosate currently has the highest global production volume of all herbicides. The largest use worldwide is in agriculture. The agricultural use of glyphosate has increased sharply since the development of crops that have been genetically modified to make them resistant to glyphosate. Glyphosate is also used in forestry, urban, and home applications. Glyphosate has been detected in the air during spraying, in water, and in food. The general population is exposed primarily through residence near sprayed areas, home use, and diet, and the level that has been observed is generally low.

For the full report see: https://www.iarc.fr/wp-content/uploads/2018/07/MonographVolume112-1.pdf

29

The Land, 'Glypho ban "Disastrous"' by Gregor Heard, Thursday 23 August 2018, page 22.

30

CropLife's press release after Josie Taylor's report: www.croplife.org.au/media/media-releases/abc-reporting-today-on-glyphosate-misleading-and-irresponsible/

31

J. Baudry et al. (2018). 'Association of frequency of organic food consumption with cancer risk: findings from the NutriNet-Santé prospective cohort study.' *JAMA Internal Medicine* 178.12: 1597-1606.

Chapter 9

32

Honey fraud report in Australia: www.abc.net.au/news/2018-09-03/capilano-and-supermarkets-accused-of-selling-fake-honey/10187628

J.-F. Cotte et al. (2004). 'Chromatographic analysis of sugars applied to the characterisation of monofloral honey.' *Analytical and Bioanalytical Chemistry* 380.4: 698-705.

CHAPTER 10

33

Henry Ford predicted in 1925, 'The fuel of the future is going to come from fruit like that sumac out by the road, or from apples, weeds, sawdust – almost anything. There is fuel in every bit of vegetable matter that can be fermented.' ('Ford Predicts Fuel from Vegetation', *New York Times*, 20 September 1925; page 24.)

34

More information on community supported agriculture farms can be found at: www.csanetworkausnz.org/

35

How we spend money is changing all the time and the government analyses these changes. For interesting data, see: www.moneysmart.gov.au

36

For the full report by Professor Alex McBratney see: http://theconversation.com/ in-100-years-time-maybe-our-food-wont-be-grown-in-soil-108049.

The International Union of Soil Scientists is also a rich resource for exciting soil news: www.iuss.org/

37

Pastures chewed down to bare earth slow down plant regrowth and after rain weeds usually proliferate. Local dust storms create poor air quality and sometimes they gather to sweep across the nation like in 2009 when Sydney turned red from so much dust. Estimates suggest 2.5 million tonnes of soil was removed in that storm alone and cost the New South Wales economy around $300M.

P. Tozer & J. Leys (2013). 'Dust storms – what do they really cost?' *The Rangeland Journal* 35.2: 131-142.

To help understand erosion and where it is happening visit the DustWatch project. It was started in 2002 and by 2005, twenty DustWatch stations, now called the Rural Quality Air Network, were set up. They're managed by Office of Environment and Heritage with 40 volunteers maintaining the monitoring sites. Hundreds of citizen scientists also send in information when dust events occur.

www.environment.nsw.gov.au/topics/land-and-soil/soil-degradation/wind-erosion/community-dustwatch

FURTHER READING

B. Pascoe (2014). *Dark Emu Black Seeds: Aboriginal Australia and the Birth of Agriculture*. Broome: Magabala Books.

W. Gammage (2011). *The Biggest Estate on Earth: How Aborigines Made Australia*. Sydney: Allen & Unwin.

T. Flannery (2010). *The Weather Makers: The History and Future Impact of Climate Change*. Melbourne: Text Publishing.

C. Massy (2017). *Call of the reed warbler: A New Agriculture – A New Earth*. Brisbane: University of Queensland Press.

P. Andrews & N. Hodda (2006). *Back From the Brink: How Australia's Landscape can be Saved*. Sydney: Australian Broadcasting Corporation.

R. Owen (2015). *The Australian Beekeeping Manual.* Wollombi: Exisle Publishing.

S. Steingraber (2010). *Living Downstream: An Ecologist's Personal Investigation of Cancer and the Environment.* Boston: Da Capo Press.

INDEX

Discover a
new favourite

Visit **penguin.com.au/readmore**